SAS RESCUE MISSIONS

FROM THE JUNGLES OF MALAYSIA TO THE IRANIAN EMBASSY SIEGE

SAS
RESCUE MISSIONS

FROM THE JUNGLES OF MALAYSIA TO THE IRANIAN EMBASSY SIEGE

BARRY DAVIES

amber
BOOKS

This Amber edition first published in 2019

First published in 1996 as *SAS Rescue* by Brown Packaging

Published by
Amber Books Ltd
United House
North Road
London
N7 9DP
United Kingdom
www.amberbooks.co.uk
Instagram: www.instagram.com/amberbooksltd
Facebook: www.facebook.com/amberbooks
Twitter: @amberbooks

ISBN: 978-1-78274-752-9

Printed in the United Kingdom

PICTURE CREDITS:
Brown Packaging: 32, 33, 35, 36, 54–55, 58–59, 60–61, 68–69, 71, 82, 98–99, 118, 121, 124–125, 128, 132–133, 136–137, 138–139, 142–143, 153, 176–177, 190
Military Picture Library: 2–3, 5
Photo Press: 8–9, 12–13, 17, 22, 74, 77, 78–79, 107, 134, 172–173, 174–175, 186–187
Private collection: 10, 11, 14, 18, 20–21, 25, 28, 29, 30–31, 38, 39, 40, 43, 44, 47, 48–49, 50–51, 52, 53, 65, 73, 80–81, 84–85, 87, 88, 89, 90–91, 92–93, 94, 96–97, 100–101, 102–103, 105, 108–109, 110–111, 112, 114–115, 116–117, 122–123, 126–127, 129, 131, 140–141, 146, 148–149, 156–157, 159, 161, 184–185
Rex Features: 26, 56–57, 62–63, 66, 108, 151, 155, 163, 164–165, 168, 169, 170, 178–179
TRH Pictures: 144, 166–167, 170–171, 180, 181, 182–183, 188–189

CONTENTS

INTRODUCTION

Fifty years after its creation, the Special Air Service (SAS) is a household name both at home and abroad. When its soldiers go into action, news of it spreads like wildfire. SAS actions are normally swift and very hard-hitting. Then, to reinforce the myth, the men of the SAS fade back into obscurity. What the public have seen of them, such as the spectacular hostage-rescue at the Iranian Embassy in London in May 1980, confirms the truth for the many other unseen actions. Yet few see the SAS for what it is: 200 men, the best our country can find, rigorously selected, highly trained and with a spirit to dare. They will go, willingly, deep behind enemy lines, take on incredible odds, and risk their lives to rescue others.

This book covers most events from the foundation of the SAS to the current war in Bosnia. The events are based around the various rescues the SAS has carried out, and relies not just on media information but true stories told by SAS soldiers themselves. The reader will discover many new, previously unseen photographs of the SAS in action, actually taken by a member of the SAS, often in the heat of battle. This gives one of the fullest pictures of the Regiment which, in our world of uncertainty and danger, has become one of the most respected and feared special forces units ever created.

TRAINING FOR RESCUE

It takes dedication, supreme stamina, iron self-discipline and thousands of hours of training to turn an ordinary soldier into a member of the SAS's crack Sabre Squadrons.

In the fields of conflict and terrorism, the one name that stands pre-eminent is that of the British Special Air Service (SAS). From its inception during World War II, this unit has developed a reputation for excellence second to none. The primary SAS roles are suppressing insurgencies and conducting daring behind-the-lines special assignments. Above all, though, the SAS is best known for its direct actions against terrorist attacks and organisations.

The SAS was originally formed as a military raiding group, designed to attack targets deep behind enemy lines. Additionally, after World War II (1939-45) it became heavily involved in intelligence-gathering and counter-insurgency operations. Towards the middle of the 1960s, life in the SAS became very boring and many soldiers, all of them individuals keen for action, left in search of adventure. About this time the Regiment started to train people in close-quarter battle techniques. Such skills had been in demand during SAS 'Keeni Meeni' (a Swahili word used to describe the sinuous movement of a snake in long grass)

Left: A soldier stops to take his bearings during a march across the windswept Brecon Beacons, the area of South Wales used by the SAS for its Selection Training courses.

Above: An SAS counter-terrorist sniper takes aim with a Finnish Tikka rifle, a weapon used by the Regiment in the 1970s.

undercover patrols in Aden. At the same time international terrorism was on the increase, with aircraft hijacking being particularly popular. So as most Western governments searched for an answer to this new threat, the SAS prepared the basis of its anti-terrorist team. Then came the Oman war (1970-76), and once again the SAS found itself doing what it does best – fighting.

However, long before the Oman war had ended, the CRW (Counter Revolutionary Warfare) cell at the SAS's UK base at Hereford became well established, partly to counter the international terrorist threat and partly to prepare the SAS for a sudden move into Northern Ireland. From CRW grew the SP (Special Patrol) team, as it was first known. The SP team grew almost overnight, and men were sent directly to the Rover factory to take the first six white Range Rovers off the production line for their transport (there is much truth in the rumours that Downing Street ordered direct action of this type in the setting up and procurement of specialist equipment). The team grew rapidly, and eventually came to be a full squadron commitment. It was a combination of the SP team work plus the CRW training for Northern Ireland that turned the SAS into an expert anti-terrorist unit. At the same time, though, it would be wrong to assume that the Regiment became inflexible. Although actions such as the Mogadishu hijack (October 1977) and the Iranian Embassy siege (May 1980) gripped the world's headlines, when the call came for British forces to enter the Falklands War (1982), the SAS was first in the queue. When the 1991 Gulf War erupted, the SAS went back to what it had started out doing in 1942: raiding behind the enemy lines.

But how are elite soldiers created? This chapter examines what the normal soldier can expect when he arrives at Hereford, and how basic Selection to join the SAS is carried out. It also covers the logical sequence of Continuation Training that takes the individual through the specialised skills of hostage-rescue work.

SAS selection is hard. The basis of the Selection system is the paramount importance of ensuring that valuable training time is spent only on the very best recruits. Candidates for the SAS need to have had at least three years' service with a parent unit to be eligible to put themselves forward. This ensures that the basic training requirements and disciplines are already in place. Selection is a funny word to use. The verb 'select' means to pick out the best or most suitable, while the adjective means chosen for excellence. The problem is that nobody picks or chooses the candidate, because as each man must earn his place it is more a case of the individual selecting himself. This is what helps to make the SAS unique as a group of individuals who can also work in a team. Candidates who pass Selection must give up all previous rank held in their parent regiment and revert to trooper. The individual must then work his way back up the promotion ladder.

Selection takes place mostly in the Brecon Beacons in South Wales, and although not a high range of mountains, they are treacherous, exposed and battered by constant weather changes. As a consequence, death from hypothermia is a constant threat, and many have suffered this fate. It is therefore essential that a diligent, self-imposed training schedule be undertaken by the candidate before his arrival at Hereford. It is also good advice not to listen to any member of his unit who has previously failed the SAS Selection course. The aspiring SAS trooper should instead concentrate on long, steady walks carrying a medium-sized rucksack. In addition, arriving at Hereford with an injury is utterly pointless – the candidate will find little sympathy from the training staff.

Selection Training adheres to tried and tested routines, and the basic format is still what it has always been – strenuous. Stamina and fitness, plus the ability to read an Ordnance

Below: SAS SP team members in the 1970s with one of their Range Rovers. The Regiment had the first six off the production line.

Right: A group of prospective recruits tramp across the Brecon Beacons during Selection – a tough test of stamina and self-discipline.

Survey map, go a long way to getting the candidate successfully through to Test Week.

Test Week comes at the end of the first phase of SAS training, and is designed to test individual fitness and stamina. It is by far the most gruelling physical requirement of the entire course. It is sound advice, if one wants to make it through Test Week, to eat as much food as possible, especially breakfast. At the end of Test Week the remaining candidates come face to face with the Endurance March. Little prepares them for this. With a rifle and bergen weighing 25kg (55lb), the candidate is expected to walk 40km (25 miles) in a time of 20 hours. This may seem an indifferent task, until one knows that the route runs up and down the Brecon Beacons.

Candidates on Test Week should always carry a full complement of clothing

The Endurance March starts early in the morning, and the candidates need to make good time by maintaining a steady overall pace by walking uphill and running downhill. Food stops should be regular – about every three hours – and rest periods best kept to 20 minutes. A brew of tea with a small amount of high-calorie food is best, for the consumption of too much food may cause candidates to vomit when they set off again. Time during the rest break is best spent checking the map and memorising the route. On summer Selection, the candidate must watch for signs of salt deficiency through sweating. It is also important to maintain a warm body temperature both on the march and while taking a break, for becoming too hot or cold severely affects individual performance and thus overall route timings. Regardless of the weather's appearance when he sets off, the candidate should always carry a full complement of clothing that will cater for all climatic conditions. The weather is likely to change, and thick fog and darkness will drastically cut visibility. The value of a good torch and several spare batteries cannot be emphasised enough – there are no street lights on the Beacons! Accurate compass work will prevent the candidate from staggering around aimlessly when visibility is reduced to arm's length. For those who can afford it, the purchase of a GPS (Global Positioning System) handset greatly

helps navigation in poor visibility, but there is no substitute for good map-reading. Exhaustion and cold can be kept at bay by sucking boiled sweets – it was for this reason that they were originally put in military ration packs.

For the lucky candidates who find the Army trucks at the final rendezvous (RV) inside the allocated time, they can rest assured that physically the worst is over. Yet there is no time for relaxation as there is still a long way to go – six months – before the candidate enters into the ranks of the SAS.

Continuation Training follows for the successful candidate, which teaches all the necessary basic skills required to be an SAS soldier, allowing him to become a new

Above: These men have just finished the escape and evasion phase of Continuation. The haggard looks are the result of a 24-hour interrogation.

squadron member. Continuation is all about going back to basics: weapon training, patrolling skills, Standard Operating Procedures (SOPs), escape and evasion exercises, parachuting, and finally six weeks of jungle training. Then, for those who cannot swim or drive, there is a crash course in both before the candidate is allowed to enter a squadron. Finally, before this is done, the successful man receives his beige beret with its famous winged dagger. As any SAS man will tell you, it is a special moment and a fabulous feeling.

There are four Sabre (fighting) squadrons in the SAS, plus R Squadron, the reserve. Each squadron is divided into four troops – Mobility, Mountain, Air and Boat Troops – and a small headquarters section. The troops are designed to operate in all terrain and environments, and thus provide the capability for all the different methods of insertion. Each patrol within a troop is made up of four men, and the four-man patrol has become the backbone of SAS soldiering. Each troop member has an individual skill, such as medicine, language, demolitions and signals. These are the basic skills, and depending on length of service it is not uncommon to find a troop member with several different skills. Additionally, each troop specialises in its own troop skills.

The SAS now runs its own Alpine training course

The Mountain Troop is responsible for all aspects of mountaineering and skiing. Training for the troop takes several forms. At times the whole troop may embark on climbing training. New members with no previous experience are taught the basics of rock climbing and abseiling techniques. In the past many SAS individuals have attended courses in Europe, mainly the German Alpine guides course. This course normally lasts one complete year, and takes two SAS personnel at a time. Those who qualify return both expert climbers and skiers. Over the past few years the mountaineering and skiing skills within the SAS have become so proficient that it now runs its own Alpine training course. The Mountain Troop is also responsible for ski instruction. Most troop members are required to instruct the other squadron members during the annual winter exercises in arctic Norway. Again several advanced ski instructor courses are available to the Regiment, both French and German. Although the SAS rarely takes part in competitions, with all this training it is not uncommon to find world-class mountaineers within each troop, and several SAS members have climbed Mount Everest.

The Boat Troop tackles all water insertion methods, including diving, boat work and even swimming ashore on a surf board. In recent years members of the SBS (Special Boat Service) have been stationed at Hereford, and cross training between the two units has taken place. Several operations have been jointly carried out using the extremely professional SBS for the water insertion element of the operation.

The Air Troop is the freefall parachute troop with each squadron, and normally its members are the 'prima donnas' of the squadron. Their role is more individual than the other three troops as they are connected with the rest of the squadron only as pathfinders.

When he has finished Selection, every SAS soldier attends the SAS parachute course. This involves four low-altitude (60m/200ft) static-line jumps, seven 'normal' (245m/800ft) jumps and two over-water jumps. Additionally, all SAS personnel practise High Altitude, Low Opening (HALO) and High Altitude, High Opening (HAHO) techniques, the latter allowing men to be dropped some 30km (19 miles) from their target. The Air Troop also covers such entry methods as microlight aircraft and power-kites.

The Land Rover 110 is outfitted with a variety of armament

The Mobility Troop is concerned with motor transport. Many missions are carried out using vehicles; indeed the SAS 'Pink Panther' Land Rover, or 'Pinkie', is known and recognised throughout the world. The Regiment decided to paint its vehicles pink when an old aircraft, shot down during World War II, was found in the middle of the desert. The sand had burnished it pink and it was difficult to spot. The current vehicle is the Land Rover 110, and this is generally outfitted with a variety of armament, including a 7.62mm General Purpose Machine Gun (GPMG), 0.5in Browning heavy machine gun, 40mm Mk 19 grenade launcher, and 30mm cannon. The vehicles are all fitted for long-range combat missions, and additional weapons can include the 81mm mortar and Milan anti-tank missile. Other vehicles used by the Mobility Troop include the KTM 350 and Honda 250 motorcycles, the Honda being preferred as it is very quiet. Courses for members of Mobility Troops cover several weeks with the REME (Royal Electrical and Mechanical Engineers) learning about basic mechanical fault-finding and repair. Training in cross-country conditions can vary from the United Arab Emirates to the deserts of America.

When a new threat is encountered, new skills are required. The need for new skills came to be appreciated with force on 5 September 1972 when, at the Munich Olympics in West Germany, eight Palestinian terrorists forced their way into the quarters of the Israeli team in the Olympic village. Calling themselves 'Black September', a name signifying the Palestine Liberation Organisation's defeat and withdrawal from Jordan in September 1970, the terrorists killed two Israelis and took nine others hostage. There was an abortive gun battle between the terrorists and the West German police trying to secure the athletes' freedom, and all nine of the Israelis died along with five of the Palestinian terrorists.

Most anti-terrorist teams have evolved in more or less the same way

The Israelis were horrified that such a thing could happen to unarmed civilians, and swiftly took action. The rest of the world was angry that the Western nations apparently had no answer to this type of violence. The West German regular police at Munich were not used to confronting dedicated terrorists armed with submachine guns. The basic lesson was not missed by many governments, and those that responded to the new threat included those of West Germany and the UK, which ordered that specialised anti-terrorist units be established immediately. Reliable information indicates that the basic agreement on hijacking was one of the main subjects of the secret talks of the G7 group of industrialised nations held just after the Munich massacre. What is certain is that it was not long after these talks that the SAS formed its own anti-terrorist unit.

For all SAS Sabre Squadrons, Close Quarter Battle (CQB) is a standard requirement whose basics are taught during the final stages of Continuation Training. The wider role of the SAS demands that each man be proficient with the pistol and the submachine gun, as well as the rifle. These skills are the very essence of being an SAS soldier. The weapon must become an extension of the trooper's hand, and feel as natural in it as a knife or fork. Some areas of operation, such as Northern Ireland or the anti-terrorist team, require more advanced CQB methods and training.

Right: The static-line parachute course for prospective SAS soldiers lasts four weeks. All must pass to wear the Winged Dagger badge.

Most anti-terrorist teams, irrespective of their country of origin, have evolved in more or less the same way. They are divided into two main groups: the assault team and the sniper team. These two elements, together with a small command and communication group, make up the unit. The strengths of an anti-terrorist team depends on tasks and/or the terrorist situation, but normally the minimum strength required is around 50 men.

The assault team focuses mainly on assault entries, concentrating mainly on all the methods of getting into the target area, be it an aeroplane, train or building. The sniper team, on the other hand, deals with any long-range threat that may present itself. Although the two teams exercise independently, a lot of cross-training necessarily goes on, thus providing the numbers to suit the situation required.

Weapons training starts from the very basics of pistol work

All members of the SAS anti-terrorist team spend hundreds of hours in the famous 'Killing House'. This is the name given to a flat-roofed block building in the grounds of the Hereford base. This structure was designed and built with the express purpose of perfecting individual shooting skills of SAS soldiers, and allows for many different sets of circumstances.

As a result of the high number of personnel practising at any one time and the amount of rounds fired daily, safety is of the utmost importance. Weapons training starts from the

very basics of pistol work, and encompasses all the problems of movement and weapons stoppages. Training then progresses to more advanced techniques using automatic weaponry.

Here the celebrated SAS ‘double tap’ is learnt. This involves firing two shots in rapid succession from the 9mm Browning High Power pistol. It takes several years to become comfortable with this method of shooting, but the results have proved themselves time and time again. Two rounds stop a terrorist far better than one, and several double taps will stop him dead. Another favourite skill honed in the ‘Killing House’ is the hostage-snatch. This is

a drill which is practised as part of the anti-terrorist/hostage-rescue scenario.

The SAS plays host at its Hereford base to an endless procession of VIPs, and many are keen to view the anti-terrorist team in action. In many ways this suits the Regiment. First of all, when not on exercise or operations, the team is permanently stationed in the camp and would be doing its normal training anyway. Secondly, all such VIPs are possible future terrorist targets, and this is a chance for the SAS to strut its stuff. The best demonstration normally involves the VIP in the hostage-snatch scenario. Those VIPs who have visited Hereford and sat in the hot seat, while black-clad figures burst into the darkened room firing live rounds at targets positioned within inches of them, will know what hell feels like.

The hostage sits in the hot seat in unqualified fright

The 'hostage' finds himself sitting in absolute darkness, surrounded by silence, not knowing what possessed him to volunteer in the first place. Abruptly, the door bursts open and a stun grenade explodes inches from him. He sits petrified as beams of laser light penetrate the blackness, searching hungrily for targets. Any VIP in the hot seat at this stage does not move. It is not that he is doing as instructed, or fearful that any movement might bring him into contact with the hail of bullets now spitting within inches of his body – it is unqualified fright. Luckily, the VIP is plucked, albeit roughly, by the black-clad figures and literally thrown out of the room, surprised later to discover there is not a single scratch upon his precious person, apart from the odd bit of singed hair!

VIPs normally arrive with an entourage of hangers-on. During the demonstration they are normally in the same room as the hostage snatch, securely trapped in the corner behind a strip of white tape. There is an excellent cartoon hanging up in the 'Killing House', which shows the boys bursting into the room and shooting the wrong way, while the instructors voice booms over the loudspeaker: 'No, no – shoot the targets.'

On average the SAS soldier is likely to expend as much as 2000 rounds of ammunition a week in the 'Killing House'. This unique facility is so sophisticated that even the Germans and Americans copied it. The various rooms can cater for scenarios ranging from several men firing from basic positions to a full room clearance, and can also emulate the internal layout of an aircraft. Despite the millions of rounds that have been fired since the 'Killing House' was constructed and the realistic conditions in which training is carried out, there has to this day only been one fatality.

(Author's Note: I knew both men involved very well. The one who died had spent many a festive Christmas dinner at my home and was a close friend. The man who accidentally shot him I also knew well after he had arrived in the squadron from the Royal Signals. After the shooting accident there were those who thought him unfit to serve in the Regiment, but I knew him better.

So heavy was the bergen he carried that he could hardly rise

I remember one occasion when we parachuted onto the island of Senya, in the far frozen reaches of northern Norway, where, in the middle of winter, the only daylight is that which sneaks under the canopy of darkness for an hour at midday. As we dropped into the blinding snow storm it was obvious that the exercise should have been cancelled, but it was too late for this revelation and there we were, knee deep in the white vomit. For four days we struggled through the blinding snow blizzard, in temperatures that fell below minus 30 degrees. Although all of us grew weaker, for the rest of the patrol it was just another 'head shed' cock-up and we would have to grin and bear it. But for the guy in question, who was my signaller and normally carried far more weight than the rest of the patrol, it was a test of human

Left: SAS jungle training in the Far East. Note the two cables used for pop-up targets – excellent for testing reaction times.

endurance. During one night as we struggled in vain to meet our RV with the rescue helicopter, the signaller collapsed into the deep snow, totally exhausted. So heavy was the bergen he carried and so thick the snow that he could hardly rise. I struggled hard, helping him to his feet, and eventually we pressed on. Despite the weight of his burden and the life-threatening conditions, he never once complained, but just gritted his teeth and soldiered on. These are the moments I recall, but the 'head-shed' do not see the consequences of such occasions when they make their wide-sweeping decisions. Some time after the death, I believe that the 'live hostage' conditions in the 'Killing House' were changed to prevent a similar episode.)

Respirators are normally carried in a container strapped to the back

All assault team members wear a black one-piece suit of flame-retardant material, and over this goes the body armour and the weaponry. The latter is normally a 9mm Heckler & Koch MP5 submachine gun that clips flush across the chest. Additionally, a low-slung Browning High Power pistol is strapped to the right leg for back-up or for use in confined spaces. In the past few years the SAS has evaluated many handguns, and the current model is the TK.

Respirators are normally carried in a container strapped to the back, although in the course of a real assault the pack is discarded and the respirator is shoved up the left arm and kept for immediate use. Most actions now involve wearing the respirator as this not only protects against gas but also presents an evil head of obscurity to the terrorists. Boots are of the non-slip type similar to professional climbing boots.

The dress of the sniper is frequently identical to that of the members of the assault team, but excellent camouflage clothing is also used. Again the same weaponry is issued, though additionally each man has two sniper rifles, one of them for daytime use and the other for night use with a special night scope. The main sniper weapon used when the author was last on the team was a Finnish Tikka M55, but this has since been changed in favour of the British Accuracy International PM sniper rifle.

The one invaluable thing that the public does not perceive is the enormous amount of

Right: Boat Troop members practise infiltration drills with an inflatable dinghy from a Royal Navy submarine off the coast of Scotland.

training that goes into creating a skilful shot, be it in the assault or sniper role. The complex shooting demanded in a hostage situation requires dedication and the ability to shoot from any position in any environment. In conditions of absolute darkness and uncertain surroundings, and to avoid shooting the wrong person, the SAS soldier must identify, confirm and act rapidly. This can only be achieved by constant and rigorous training in realistic conditions. Abseiling down the side of a building in full gear is not as easy as it sounds (at the Iranian Embassy one of the SAS men got caught up in his rope). The entanglement of ropes, weapons and stun grenades has to be cleared rapidly to allow the soldier to close with the terrorists.

Above: Freefall parachuting is an integral part of SAS Air Troop tactics. The men learn such skills after they join their respective squadrons.

Of course, the specialist equipment does not just appear out of thin air on SAS demand. It is normally developed for a specific purpose and can take years to perfect. During the Iranian Embassy siege, for example, one can observe the group of black-clad figures balancing precariously on the balcony attaching something to the window. Seconds later they slip a few feet to the side and, with a loud 'boom', the window is open and in jump the beasts from hell, lobbing stun grenades before them. It is not always that easy. To cut a hole in a wall or take out a window frame requires a precise explosive cutting charge. The guidelines are simple: make a hole in the wall or window but do not under any circumstances damage any hostages inside. In order to perfect this technique, the Ministry of Defence very kindly allowed the SAS to train on a number of old married quarters, most of them tucked away discreetly in some remote corner of a military barracks or airfield.

(Author's Note: In the early days I did my 16-week demolition course, and very good it was. We learnt all the technical formulas for applying just the right amount of explosive to cut steel or concrete. But above all I learnt the basic SAS formula for estimating the correct amount if all else fails – it is called 'add P for plenty'. Once when we were assessing the amount of explosive needed to cut through a normal house wall, several of us strapped a frame charge to the side of a house. The house had been borrowed for the occasion from the RAF, and luckily it stood with two others out of sight. Having placed the charge firmly on the outside wall I rigged the detonator to the 'Shrike', which was a device specially developed for the SAS to set off explosives. It was decided that we would carry out a practice house assault on the building by entering through the hole in the wall that we were about to make. So, poised safely around the corner of one of the other houses, I set the charge off. 'Go, Go, Go,' I shouted. Through the dust and the debris we charged forward, only to find no hole – in fact no house. The house had disappeared, too much explosive by the looks of it. Oh well, back to the drawing board.)

Most SP team personnel will take part in several exercises

One whole squadron is committed full time to the SAS anti-terrorist team. The squadron is normally broken down into the A and B teams. One remains in Hereford, normally practising in the 'Killing House' and doing internal courses of study, while the other is generally taking part in terrorist scenarios to duplicate the various types of hostage situation. At some stage during the four-month period most SP team personnel will take part in several exercises, covering trains, buses, aircraft, ships and buildings.

During any major terrorist incident in the UK involving the SAS, Hereford is normally warned to stand by, via the excellent network that exists between the chief constables and the SAS. Control of all terrorist incidents in the UK is firmly in the hands of the civilian authority. Even with the SAS on site, the Regiment acts only when the situation demands the use of immediate action to stop the loss of life, and then only when command has been officially passed from the police to the military. That is when it gets really exciting.

Operations overseas depend totally on the country and the situation, at such times. However, it is useless relying on the Foreign Office for help – any SAS man worth his salt just gets in there and does what is required. A perfect example of this was when two SAS personnel went to put down a coup in the Gambia (see Rescuing VIPs, pp140-165) instigated by Libya. While President Jawara was out of the country attending the wedding of Prince Charles and Lady Diana Spencer, the rebels seized the capital, taking the president's wife and family hostage. The SAS lads got as far as neighbouring Senegal, where the Foreign Office advised them strongly not to go. Not only did they go, they personally rescued the president's family and played a major part in putting down the rebellion.

Apart from the 'Killing House', the SAS has an enormous variety of equipment and, in the region of Hereford, many training aids, most of them in constant use. Several aircraft types are available to practise both aeroplane entry and assault. There is also access to a complete two-storey building know affectionately as the 'Embassy'. Railway carriages are positioned for the construction of train hijack scenarios, and there are plenty of roads to do coach and vehicle ambushes. Each move is practised and re-practised over and over again, and this process contributes enormously to the success of the SAS. As the Regiment will confirm, there is no substitute for hard work, constant practice and attention to detail.

The SAS discovered the simple principle of having two ladders side by side

Not all the skills are devised by the SAS – many are developed by the anti-terrorist units of other countries. Since 1977 there has been constant cross-training between many of the world's elite forces. From this type of cooperation comes the birth of new ideas and methods. During the Mogadishu hijack, for example, the SAS discovered the simple principle of having two ladders side by side. This allows one man to open an aircraft door, while the second man is free to react the moment the door is clear.

Depending on the country, the specialist teams for the new counter-terrorist role were formed from either existing military or police units. In West Germany, for example, the anti-terrorist unit was developed from the border police, *Grenzschützgruppe* 9 (GSG 9). Similarly, in France the *Groupement d'Intervention de la Gendarmerie Nationale*, (GIGN) is also a police unit. In the USA, on the other hand, Delta Force was formed from American special forces units in a fashion similar to that of the British SAS.

The origins of the counter-terrorist force in the police or military may not seem to be significant, but in reality it is. The SAS work on a rota system. The rota is broken down into roughly five-month blocks in the form of five months in the Middle East or wherever the fighting is at the time, five months on the anti-terrorist team, five months on matters the author cannot reveal, and five months training on new techniques and equipment. The point is that the SAS is always active, and its soldiers have a lot of real combat experience.

Sayeret Matkal comprises soldiers of very high calibre

The German GSG 9 and French GIGN, on the other hand, only have their police duties. That said, this type of origin does give them a good edge during domestic terrorist instances. Until Mogadishu, GSG 9 was untested. Conceived after the fiasco which ended with the Palestinian terrorist attack on the Israeli team during the 1972 Munich Olympics, GSG 9 saw service under operational conditions during the Mogadishu episode (the first time a German military-style action took place outside Germany since World War II). It performed brilliantly under the inspired leadership of Ulrich Wegener.

The Israelis would have us believe that their anti-terrorist force is an extemporised unit created as and when required. This is not the case, for the Israelis use a force called *Sayeret Matkal* (The Unit) comprising soldiers of very high calibre, most of them from the parachute

Right: A column of SAS 'Pink Panther' Land Rovers photographed in Saudi Arabia. Vehicle skills are the preserve of Mobility Troops. Note the smoke dischargers on the front and rear.

regiments of the famous Golani Brigade. The author has never worked with them, but has trained with them and they are very impressive. Again the unit was first formed after the Munich Olympic massacre. Later it carried out the fantastic hostage rescue at Entebbe in July 1976, and is now frequently used as the strong arm of Mossad, the Israeli intelligence service.

The terrorists shot and killed two hostages: an Algerian and a Vietnamese

Although the French GIGN had participated in several successful actions, it had to wait until 1994 to get its main chance. (Author's Note: I first met the French *Groupement d'Intervention de la Gendarmerie Nationale* in its early days. It came to Hereford when I was on the team, and I was not greatly impressed. The man in charge of the GIGN was Capitaine Christian Prouteau, a tall, good looking officer.) The author watched with interest the first television news of a hijack that had taken place at Algiers airport. The news reports indicated that an Air France Airbus had been boarded by four Islamic fundamentalists, who had taken the 220 passengers and 12 crew hostage. Shortly after this they increased their profile by releasing 19 people, mainly woman and children. However, this act of mercy was soon overshadowed when the terrorists shot and killed two hostages: an Algerian policeman and a Vietnamese diplomat. Hours later, early on Christmas morning, they released more women and children, bringing the total of released persons to 63. A few hours later, the hijackers shot a third hostage, a 27-year-old Frenchman, who was a cook at the French embassy in Algiers. Soon after this, the airliner took off, landing early next morning in Marseilles.

What the general public did not realise is that the Air France airliner was one of the worst hijack scenarios possible, and required an IA (Immediate Action) response. This is carried out

10 FG 68

Above: French GIGN members photographed after their stunning hostage-rescue operation at Marseilles in December 1994.

only as a last resort when people are being killed. There is nothing sophisticated about the assault. The GIGN men had to get on board the airliner as soon as possible to neutralise the terrorists and free the hostages. The risks to both the assaulting force and the hostages is very high as a result of the lack of surprise or diversion.

After the airliner had landed at Marseilles, the airliner was immediately surrounded by snipers of the GIGN. Then, at 15:45 hours, the airliner moved towards the control tower without permission. The demands from the authorities for the aircraft to stop, and then to move no farther, were met with several shots from the terrorists at the control tower. By this time the GIGN commander must have realised the terrorists' intentions and ordered the IA to be carried out. By 17:17 hours, GIGN assault units could be seen racing towards the rear of the aircraft, and sniper fire could be heard. Denis Favier and Olivier Kim, the team commander and his second-in-command, stormed the starboard-side forward door. They used normal airport landing steps to gain entry, and despite trouble with the door the entry was very slick considering the conditions. At the same time Capitaine Tardy led another unit via the starboard-side rear door, using the same entry method. Once on board both groups made their way towards the cockpit, where most of the terrorists had gathered. A third GIGN group moved into position under the belly of the aircraft, ready to receive the hostages. Inside the Airbus the two teams moved quickly, separating the terrorists from the hostages and disembarking the latter as soon as possible.

By 17:39 hours the four terrorists lay dead. Although the hostages got away with a few minor cuts and bruises, the cost to GIGN and the aircrew was heavy. Nine GIGN men were wounded, two of them quite severely (one lost two fingers of his hand and one was shot in the foot). The three members of the crew who were hurt had unfortunately been trapped with the terrorists in the confined cockpit area (during the assault the co-pilot managed to climb out of a small window and throw himself to the ground, and despite a broken leg managed to effect his escape).

Colonel Beckwith, the founder of Delta Force, had been attached to the SAS

Later reports indicated that the terrorist plan was to fill the aircraft full of fuel and blow it up over Paris, and this theory became more plausible when 20 sticks of dynamite were found under the front and mid-section seats. The author now believes that many Parisians may count themselves lucky that their country has a force such as the GIGN.

Delta Force was raised in 1977 by Colonel Charles Beckwith. The author knew 'Charlie' Beckwith and he was some soldier. He had been attached to the SAS on exchange in 1962, and had served in Malaya. When he finally returned to the USA the first thing he did was write a report that advocated the creation of a special unit based along the lines of the SAS. It then took Beckwith until 1977 to get his plans for Delta Force passed by the Joint Chiefs of Staff.

Delta Force's first hostage-rescue operation ended in total disaster. It started with the 1979 coup in Iran, where the Shah was toppled from power and replaced by the Ayatollah Khomeini. The army then effectively disintegrated, to be partially replaced by a rabble known as the Iranian Revolutionary Guards. These militants took over the American Embassy in Tehran and held over 100 Americans hostage. Delta Force was sent to rescue them. It was one misfortune after another, as described below, which goes to prove you cannot win them all.

Saudi Arabia has one of the best anti-terrorist teams in the Middle East, this team having received training from both the British SAS and German GSG 9. The SAS training team was first involved in 1978, when a six-man team arrived in Jeddah. A preliminary reconnaissance had been carried out by an Arabic-speaking major one month before the arrival of the main team. Both the SAS and the students, all selected from the palace guard, were housed in a barracks on the western edge of the city, close to the old disused Royal Palace grounds. These grounds became the primary training area.

The SAS team used a training programme similar to its own, starting with the basics of weapons training and working up to full-scale attacks on both buildings and aircraft. The daily programme was adjusted to accommodate both the midday heat and religious prayer. Morning classes would normally start with physical exercises at 05:00 hours followed by classroom lessons. Breakfast was taken around 09:00 hours, with lunch at 13:00 hours. There would then follow a sleep break until lessons resumed between 17:00 and 19:00 hours in the evening. Classroom lessons covered dry weapons training, including the Browning High Power and Heckler & Koch MP5.

The training area allocated to the SAS was the grounds of the old palace

In week two, range work began. Special ranges had been constructed in the desert, both to cover normal CQB shooting and 'Killing House' techniques. In the case of the latter, the walls of the building were constructed out of Hessian cloth, secured behind a sand bank. Safety was paramount, so command and control on the ranges required several team members at once. The students progressed from basic pistol work, firing double taps, to short,

Left: Men of the Saudi Arabian anti-terrorist team with one of the SAS men sent from Hereford to teach them hostage-rescue tactics.

Above: Putting theory into practice. Members of the Saudi anti-terrorist team carrying out vehicle ambush drills on a range near Jeddah.

controlled bursts with the MP5. As the students moved on to the makeshift killing house, they were formed into teams, learning the more difficult house-clearing techniques.

As previously mentioned, the training area allocated to the SAS was the grounds of the old Royal Palace, which contained about 300 houses of different styles and designs, laid out in a variety of street patterns. The entire complex was surrounded by a 3m (10ft) wall, which afforded excellent security. This proved to be a major asset in training the Saudi anti-terrorist team, especially in house-clearing and vehicle ambush drills. Given that the SAS were allowed to blow off doors and generally wreck a building before moving on to another, it added a great deal of authenticity to the assaults.

Towards the end of the training, with several members of the Saudi royal family present, a full-scale exercise was carried out both on the buildings and in the streets.

Aircraft assaults were practised at the airport. At first a Lockheed C-130 Hercules military transport aircraft was used – at the time the Saudi royal household used a converted C-130 for its VIP transport. An accident happened during one practice assault at the rear door of the C-130. When the door was opened an SAS instructor had one of his thumbs completely severed. Fortunately the limb was saved by a doctor who sewed it back on. Later the students practised on a Boeing Model 727 airliner, and again a final exercise included a full-scale assault on the aeroplane itself. Less than a year later the Saudi unit went into operation for real when religious fanatics took over the main mosque in Mecca. Thanks to its training it did a great job.

RESCUING OUR ALLIES

The SAS has proved invaluable with regard to saving Britain's, and the West's, allies in times of crisis. Both in Malaya in the 1950s and Oman in the 1970s the Regiment's men helped turn the scales against communist insurgents.

It was in the role of rescuing Britain's allies that the modern-day 22 SAS was born. The Malayan 'Emergency', which began in 1948, was a conflict between the British administration in Malaya and the mainly Chinese Malayan Races Liberation Army (MLRA). In 1950, a British force was raised by Lieutenant-Colonel 'Mike' Calvert to combat the MLRA, or Communist Terrorists (Cs) as they were known. This early unit also formed the basis of the four-man patrol, and proved that it was feasible to undertake protracted operations under very hostile and difficult conditions. Such arduous tasks, all performed in actual or potential enemy territory, produced tough, self-reliant men who were capable of living in the jungle alongside the local population. By the mid-1950s the unit, known as the Malayan Scouts, provided the catalyst that led to the creation of 22 SAS.

At this time entry into the jungle was normally by parachute, or tree jumping as it was known. This technique

Left: An Omani UH-1 helicopter carrying SAS troops comes into land during Operation 'Jaguar' in Dhofar, Oman, in October 1971.

was at best very dangerous, and even those who survived the initial impact onto the jungle canopy still had the further danger of lowering themselves to the jungle floor. With the increasing usage of helicopters for team insertions, the practice of tree jumping declined.

From the early days of the Malayan Scouts the principles of SAS Standard Operating Procedures (SOPs) were laid down. These included communication in whispers or sign language, and strict segregation of cooking and sleeping. To get a good night's sleep the hammock was adopted. At first this was fashioned from parachute material, but today a purpose-made hammock is issued to troopers. In the jungle one is permanently wet, either from sweat or rain, so it is a good idea to have one set of clothing for day wear and one dry set for sleeping. Changing from wet to dry, under the protection of one's poncho, is a real luxury,

Left: 'Mad' Mike Calvert, whose ideas on long-range jungle patrols led to the creation of the Malayan Scouts.

but putting on cold, damp clothes the next morning is a real drag. Likewise in the 1950s, the problem of carrying ammunition, medical equipment and radios (water is in abundance in the tropics) curtailed another luxury – food. It was impractical to carry 14 days' rations in conjunction with the rest of the kit, so before departure the ration packs had to be stripped down to the minimum weight. And going back to SOPs, all litter would be carried back to camp. It was this experience that eventually led to the development of the British ration pack.

The Malayan jungle also produced another hazard: health, or rather its rapid deterioration. Apart from the leeches, hornets, mosquitoes, ants and snakes, a host of deadly diseases lay in wait for SAS patrols. A fallen tree in the middle of the night can, and indeed has, produced several SAS casualties. In the dark jungle medical aid has to be provided by a patrol member, and thus the need for the SAS medic emerged. This had two benefits: the medic was capable of dealing with most emergencies, and even minor surgery. His field pack would contain a wide variety of drugs, dressings and surgical instruments, but most of all it would contain a small red book entitled *The Traveller's Guide to Health*. The second benefit was the use of the patrol medic in the 'hearts and minds' role. The provision of medical facilities is one of the best ways for an SAS patrol to gain the trust of the local population.

Below: A group of Malayan Scouts take a breather whilst on operation in early 1951. Their training base was at Johore, near Singapore.

It openly shows that you care, and that you are concerned. Moreover, many of the indigenous people treated by SAS medics, such as in Malaya in the 1950s, have not had the benefit of modern-day drugs, thus the effect of administering penicillin can have almost magical results.

SAS patrols in Malaya operated in the dark,dank jungle, and almost always in enemy territory. Thus the Malayan campaign was of great importance to the Regiment, as it laid the principles on which the SAS was based: Selection Training, SOPs and above all the four-man basic unit.

The war in Oman offered a challenge that the Regiment eagerly accepted

The war in Oman (1970-76) must stand out as *the* classic SAS operation. It was tailor-made to test the basic regimental skills that sets the SAS apart from the rest. The war's importance, both military and political, had far-reaching effects on the whole of the Middle East. Moreover, the success of the SAS and the Sultan of Oman's forces helped stabilise the attitude of many neighbouring countries, pushing back the tide of communism that had threatened to engulf the rich oilfields on which the West depends. It offered a challenge that the Regiment eagerly accepted, and turned out to be a classic counter-insurgency campaign of modern times. It was also a war the SAS desperately needed. The period of relative peace that had reigned since the end of the troubles in Aden (1964-67) had not benefitted the Regiment much. Despite the extensive training activities embarked upon by the SAS squadrons, many of the younger members had started to leave, tempted at the time to be highly paid bodyguards or by more active security jobs. Unfortunately, this minor exodus took many of the best NCOs, although some of them did return once the Oman war was under way.

The size of the Omani conflict was small, yet it combined all the elements of modern warfare. Navy, air force and army were all combined with one goal: to win. It cannot be said that the SAS won this war on its own, for much of the fighting was done by the Sultan of Oman's own forces. But undeniably the one thing the SAS did do was bond together the *firqats* (Dhofari irregulars formed from former insurgents into companies by SAS units, which operated in the country under the guise of British Army Training Teams). These *firqats* and the SAS went on to become the lead elements in most battles in the early days. It was the trust between Dhofaris and the SAS that won the Dhofar war.

The SAS had served in Oman previously during the Jebel Akhdar campaign (1958-59), and now it was back. In the 1970s the war took place in the south, on a huge mountain massif known as the Jebel Dhofar (the repressive regime of Sultan Said bin Taimur had goaded the Dhofaris into rebellion). It was a strange refuge for wild tribes and 'freedom fighters'. In summer it is a place of great beauty, where lush green grass blows in the cooling winds, and trees give homes to birds and other small animals. The problem was the People's Front for the Liberation of the Occupied Arabian Gulf (PFLOAG), known to the men of the SAS as *adoo* (Arabic word for enemy), which occupied most of the Dhofar region with the exception of the coastal towns of Salalah, Taqa and Mirbat. The rebels roamed free over the Jebel Dhofar, eating further into the beleaguered areas around Salalah. The base used by the Royal Air Force just north of Salalah was itself virtually under siege, and without outside intervention the Sultan's small defence force was on the way to defeat.

Within weeks, men of the SAS had been sent to Oman

Although not a military man, the old Sultan had decided to send his only son Qaboos to the Royal Military Academy at Sandhurst, where he was commissioned into a British regiment. His stay in England was far from wasted, for with personal skill and the encouragement of his friends he had observed the workings of various councils and committees, and in general had familiarised himself with the intricacies of a modern state. His return home had not been a

joyous one, though. The young Sultan could see the plight of his country and argued for change. His father's answer had been to further restrict his son's movements, and accuse him of becoming a 'Westerner'. However, the young Qaboos bided his time. His chance came on 23 July 1970, when a palace coup took place. Aided by the young Sheikh Baraik Bin Hamood, Qaboos deposed his father in a bloodless takeover. Within weeks, men of the SAS had been sent to Oman to provide the new ruler with advice and assistance.

Above: Malayan Scouts near Ipoh, late 1950. Their main tasks were to set ambushes and direct RAF strikes against guerrilla positions.

For some time Colonel 'Johnny' Watts, the commander of 22 SAS, and his second-in-command, Major Peter de la Billière, had been making preparations behind the scenes, and

Left: A different jungle war – Borneo (1963-66). The lessons that had been learnt in Malaya were put to good use against the Indonesians.

several secret reconnaissances had been undertaken by men of the Regiment. For the average SAS soldier, who was getting on with yet more training, the Oman war came as a bit of a shock, but was nevertheless a welcome break from base routine. For the SAS, the first action started in the northern Oman peninsula, at the very entrance to the Straits of Hormuz.

The arms shipment was to be accompanied by a group of Iraqis

The author takes up the story: 'I was walking across the square from the direction of the "Kremlin" [the SAS Operations, Planning and Intelligence cell at Hereford] when a voice called out to me, and I turned to see a colleague running in my direction. Like me he was dressed in a suit, as many of us were due to attend the marriage of a squadron member in a few hours' time. "Get your kit, we are going." He didn't stop, but ran on past.

"Where to," I shouted at his receding back.

"Middle East," was all I heard before I too broke into a run, heading directly for the accommodation basha. There were a few guys still getting dressed for the wedding, but the bulk had already left for a few beers before the ceremony started. The sudden appearance of the squadron commander soon made me realise that this was not a drill. Quickly he informed those present on the latest situation. It would seem that some Intelligence guy who had been lying low in a village off the Musandan Peninsula had picked up information on a huge arms shipment due to arrive at the tiny coastal village of Jumla. This arms shipment was to be accompanied by a group of Iraqis with communist persuasions. According to British Intelligence, this was translated as the main communist head-shed moving into the area to take advantage of the turmoil within Oman. But the SAS was going to snatch them and nip the whole thing in the bud.

'A few hours later, with most of the squadron still dressed in suits, we made our way to RAF Brize Norton. A day later, after a short stop in Cyprus, we landed at the British airbase in Sharjah. Without leaving the confines of the airport, we loaded our bergens and weapons onto several three-ton trucks and were driven out into the desert. To all intents and purposes this was done for security; in reality the whole squadron was dumped in the middle of nowhere. Yes, still dressed in our suits! Like all good SAS men, we dived into our bergens and changed into combat gear. A few of us then used the spartan scrub to put up some bashas for shelter, before making the inevitable brew of tea. We waited. Later that evening several choppers came in and we were ferried over to a camp site on the eastern coast. This was done for several reasons. First, the area was similar to the one we would attack, although it was minus a village. Second, a Royal Navy minesweeper, complete with several rigid raiders manned by the SBS, was due to arrive off our coastal campsite.

It was suspected that the enemy had received wind of our assault

'The plan called for the SAS to be inserted by the SBS onto the small beaches both north and south of Jumla. This would be done during the hours of darkness and would allow enough time for us to scale the high rocky peaks that towered out of the sea and surrounded the entire village. By dawn, with the SAS in place to stop any enemy running away, a combined force from local Arab states would assault the village and capture both the Iraqis and the equipment. On the surface a sound plan: not only would it protect the Straits of Hormuz, through which half the world's oil passes, but it would stop a major catastrophe in the area. But the reality was rather different.

'The night before we left our training area to board the minesweeper, a priority signal arrived. By the light of our camp fire, Major Alastair Morrison read it to us. We listened with dread to his solemn words. It would seem that the Intelligence guy, who we assumed was still in the village and still passing back vital information, had seen a ship docking at the small village port. He reported that heavy machine guns and mortars had been unloaded, together with several large boxes of mines. It was suspected that the enemy had received wind of our assault and planned to oppose it. The final words of the message were chilling: "expect to take up to 50 per cent casualties." Silence fell on the not-so-happy group. Why do they have to tell you things like that the night before a mission?

'As if being blown to smithereens on the beach wasn't enough, the whole plan was delayed for 24 hours, but only after we had put to sea. The reason was bad weather. When the time came to leave the minesweeper and load our sea-sick bodies into the Rigid Raiders, I would have happily taken on the enemy single-handed rather than stay a second longer on that ship. However, as we approached the beach in pitch darkness all thoughts of sea sickness diminished, and I concentrated on looking through the night scope. There was nothing to see other than a few faint lights coming from the village. We hit the beach at full speed and I was actually thrown from the Raider in what looked like a spectacular, head-long dive. Somehow I landed with a perfect para-roll and came up on my feet. Several of the guys said the dive looked real gung-ho (I wanted to kill the stupid SBS boat handler!).

'Almost instantly we started to climb. It was steep and hard going, but Mountain Troop led the way, choosing the easiest route. An hour before dawn we were in position, overlooking the village to the open sea beyond, our only casualty a broken finger. I should mention at this stage that some bright spark had come up with the idea of giving every SAS man at least six *shemaghs* (Arab headdresses) of various colours and design. The thinking behind this was simple. The assaulting Arab force was made up of several different local forces, all with a different coloured *shemagh*. To avoid any possibility that we might be mistaken for the enemy, as the different units approached our positions we were to change headdress to

Below: SAS soldiers being transported by ship to make a night landing on the Musandan Peninsula, Oman.

Above: Members of the author's squadron prepare for their assault on Gumla – which unfortunately was the wrong village!

comply with theirs. In reality, 40 SAS soldiers were all converted into Tommy Cooper look-alikes as we frequently dived into our bergens looking for the correct *shemagh*.

'The light rapidly improved and suddenly, about a mile out to sea, we spotted the landing force. It was heading at top speed for the beach. I have to say that from our position, high above the village, it looked impressive. The landing craft hit the beach and several shots rang out. This is it – here we go. Then slowly the villagers started coming out of their homes to greet the soldiers (visitors were rare in this remote area); tea and coffee were distributed and everything seemed very amiable. Then one of the white officers got tough: "OK. Where are the Iraqi rebels? We know they are here in Jumla."

'"If they are in Jumla, why are you here in Gumla?" came the surprised reply. We had invaded the wrong village! All that hype and planning gone to pot. There were no more than half a dozen villages along this hostile coastline, but we picked the wrong one. By the time the head-shed had confirmed the fact, and sent us racing across the jebel top, it was far too late. The rebels had long gone, and so had most of their equipment.

'We were all pulled back to the British base at Sharjah, where we sat around while a new plan was hatched. Apparently the enemy had taken refuge in a stronghold called the Wadi Rawdah. Even my best effort at explaining to anyone what this place looked like would not do it justice. It was a mighty bowl within the jebel structure. The sheer rocky walls towered up to 305m [1000ft] even at the lowest point. On the seaward side there occurred a natural, narrow split in the rock structure, which allowed for entry and exit. The whole valley was home to a strange tribe called the Bani Shihoo. Reportedly a vicious people who had rarely seen a white man, their main weapon,

Above: A Royal Navy minesweeper lying off the village of Gumla, following the beach assault, as described by the author in this chapter.

apart from the odd rifle, was a vicious-looking axe. Like most rumours it was false.

'The operation restarted with a parachute night drop carried out by two freefall troops. I think I am right in saying it was to be the first operational freefall drop the SAS Regiment had ever undertaken. The rest of us would go in by chopper as soon as the boys had secured a landing position. I had time to go to the local base NAAFI for a drink, but just before I left I gave "Rip" Reddy, one of the freefallers, a hand with his kit. He was fairly new to the squadron, but was a great guy with a real sense of humour. He joked about the amount of kit he was dropping with. The aircraft took off around 03:00 hours, and an hour later the men jumped from a height of 3350m [11,000ft]. Paul "Rip" Reddy was killed. Of all the deaths in the Regiment, this one stunned me. To this day I can recall wishing him luck before I went off for a beer. He was so young, fit and full of life, but that would sum up most men who have died in the SAS.

'Before the news broke, the rest of the squadron had been trucked to a camp site closer to the wadi. Most of us were still sleeping when "Boss" Morrison came running round, shaking us awake and telling us to grab our kit and get on the chopper. He never said that "Rip" was dead, but the speed of our move said something was up. This was also the first time I had seen a load master sitting in the door of the chopper manning a fixed GPMG. Arriving from the air, the sight of the Wadi Rawdah takes your breath away. But we had more pressing things on our minds than taking in the spectacular view. As the chopper touched down we all jumped clear, fanning out in a defensive arc. That's when I saw two of Freefall Troop bring a body bag forward, and place it on the chopper. Paul Reddy was a good man.

'We stayed in the area for about two months, using Sharjah as a base. I was pulled off to do psychological operations, an interesting task that gave me a good insight into the northern part of Oman. One of my tasks was to pre-visit the small towns and villages prior to the arrival of the new Sultan. I would dispense a radio to most adults (a new Omani broadcasting service was being set up) and give T-shirts and flags to the kids. One of my visits took me back to the Wadi Rawdah, where I came face to face with the Bani Shihoo tribe. They were a strange community. They built small houses out of rocks that had been fashioned into uniform blocks and many had a pitched roof. Due to the lack of water and the rock structure of the wadi, they had carved out huge cisterns to collect rainwater in. Their survival in this inhospitable area (I could see little sign of vegetation, and no animals) is a lesson to the human race in man's ability to adapt. Eventually, we went home, but only for a short while.'

The SAS Headquarters was established at Um al Gwarif

After the abortive raid in the north, the war in Oman continued, but this time in the south. After the first initial reconnaissances by Colonel Watts and Major de la Billière, it was decided to occupy the southern coastal towns. The SAS HQ was established at Um al Gwarif, on the outskirts of Salalah, the southern capital, and just a short distance from the RAF base. While Salalah and the air base took several long-range attacks, both were defended by an outer ring of fire bases called 'hedgehogs'. These were situated in a defensive arc between the Jebel Dhofar and the air base, and were manned by the RAF Regiment. They were well equipped with mortars and Green Archer (radar that detected and back-tracked enemy artillery fire to locate the guns), and this ensured that the *adoo* could not bring their heavy weapons in too close. The more distant towns of Taqa and Mirbat, which were manned by the SAS, were open to the full brunt of the *adoo*. In the days before the war was taken onto the Jebel Dhofar, these locations came under constant attack. Hardly a night would go by without Taqa or Mirbat coming under fire.

Daily life evolved into training the *firqat* and defending the village

'Life for the SAS training teams, before taking the war up onto the Jebel Dhofar, was confined to village life and raising the *firqats*. Taqa, a small coastal village midway between Salalah and Mirbat, was the base and tribal home of the *Firqat Kalid bin Walid*. This *firqat* eventually mustered more than 80 men. They were also fearsome fighters. During the old Sultan's frugal regime many young men had left the village and travelled widely within the Arab world. Fortunately for the SAS, some had taken training with the Trucial Oman Scouts and had a very good knowledge of British military tactics. It was common to see the section commanders give the same hand signals that could be seen on the training areas of the Brecon Beacons.

The SAS normally lived in the village with the *firqat*, and in the case of the Taqa they occupied a two-storey building overlooking the village square. Daily life for the SAS evolved into a number of days training the *firqat* and a number of days defending the village. The village was a collection of mud huts that seemed to have stood timeless for a thousand years. Some 200m (660ft) to the south was the sea, and to the north was the Jebel Dhofar. The bulk of the village sheltered under a small escarpment about 15m (50ft) high, on top of which stood an old Beau-Geste-type fort. A troop of Baluchistani soldiers occupied the fort, which in turn was surrounded by a razor-wire fence in a half-moon shape. To the left of the fort was a mortar pit, with an artillery piece situated to the right.

During the day two mortar-trained SAS men would sit waiting patiently for the first distant plop. This first indication of an *adoo* attack would normally bring the customary first two mortar rounds down the tube in reply. It was a kind of game. By day the attack would normally be from a big gun fired from the tree-lined escarpment that lined the Jebel Aram and looked out over Taqa. The gun was so well hidden that the BAC Strikemaster light attack aircraft of the Sultan of Oman's Air Force could not locate it. The SAS, too, could do little to alleviate the situation. On one occasion, for example, several shells landed in the village, killing one woman and wounding several others. But the *adoo* did not always have matters their own way. Out-ranged by the gun, the mortar had its range boosted by pouring 5cm (2in) of petrol down the barrel. This increased the distance dramatically, but is not a practice to be recommended as, firstly, the extra range is not constant and, secondly, one needs to keep a careful eye on the bottom of the tube and base plate for the moment that cracks start to appear. Nevertheless, the boosted mortar gave the *adoo* something to think about!

During the monsoon season the *adoo* made several close-range attacks

Most of the attacks came at night, when a stand-off battle of some 2000m (6555ft) would be exchanged. These normally lasted for about 10 minutes, initiated by the *adoo* mortars backed up with small arms. During the monsoon season the *adoo* made several daring close-range attacks, managing to get within spitting distance of the wire perimeter before being repelled. There is some evidence to suggest that these close-in attacks were training for the assault on Mirbat that came later.

On one occasion at Taqa the men in the mortar pit heard noises inside the wire compound and quickly asked the fort to have a quick look with the night sight. As the mortar pit was isolated from the fort, the ground to the front and open side had been festooned with razor wire about a metre above the ground as a means of stopping any sudden rush by the enemy. The fort reported seeing nothing, and things settled down for a few minutes, until a sudden fall of stones could be heard just a few metres away. Two grenades were quickly thrown in that direction, and the fort, from its elevated position, laid down several protective bursts of heavy fire across the pit. At this stage the two mortar men, who normally slept in the pit, made a run for the back door of the fort. Nothing else was heard, but next day a hole was found in the wire and a dead donkey lay 20m (66ft) away with boxes of Soviet-made ammunition scattered around its body. The position of the mortar pit was comparable to that of the gun pit at Mirbat – the prime target during that engagement.

SAS mortar teams would sit with two bombs at the ready

Most of this fighting was stand-off stuff that lasted no more than 20 minutes, and normally consisted of mortar and light machine-gun fire. This tactic became so routine that the SAS teams would be waiting, trying to anticipate from which direction the attack would come. SAS mortar teams would sit with two bombs at the ready. As soon as the 'plop plop' of the enemy mortars were heard, the SAS would immediately dispatch retaliatory rounds. The exchange was normally followed by small-arms fire. In addition, the SAS would do night penetration patrols and lay unmanned demolition ambushes.

By day the SAS men would recruit and train the local male population into *firqats*. These indigenous troops were to bear the brunt of the *adoo* assault when the SAS and Omani forces finally attacked the jebel. As it turned out, some *firqat* units were better than others. The *firqat* from Taqa, the *Kalid bin Walid*, was superb. Many of the men had already spent time in

Right: The formidable Wadi Rawdah, home of the Bani Shihoo tribe, which the author met in the summer of 1971.

Above: An Omani 25-pounder brought onto the Jebel Dhofar to support Operation 'Jaguar' in October 1971.

some military unit, such as the Trucial Oman Scouts, and were well schooled in basic tactics. Some had been *adoo,* but had been alienated by the hard-line ideology of the PFLOAG and had seen the work and determination of the new Sultan, and therefore returned to play a vital part in building their country. Once the *firqats* were ready, the time came to attack the *adoo* in their own back yard, and plans were accordingly made to assault the Jebel Dhofar.

Operation 'Jaguar' was launched to establish a firm base on the Jebel Dhofar, and it began in October 1971. Dispensation had been given by the senior Qadi (religious leader) to all Arabs fighting during the Ramadan period, which is normally a time of fasting. This was the start of the war proper: almost two full squadrons of SAS, together with their *firqats*, spearheaded the operation. Additionally, several companies of the Sultan of Oman's Armed Forces and various support units also took part. The whole force was led by Johnny Watts, a brilliant commander who above all had the respect of his men. Watts had a quick, decisive mind, yet he would not commit the SAS without committing himself. He was no stranger to the battle front, and often could be seen running forward with one gun group or another carrying boxes of ammunition, shouting orders as he did so.

Getting the men onto the Jebel Dhofar was not easy, even though a diversionary plan had been implemented several weeks earlier (heavy patrolling had been initiated from Taqa and

Mirbat in the direction of the Wadi Darbat, which had always been an *adoo* stronghold, with the intention of making it seem that a full-scale attack was imminent). Helicopter hours were limited, and although some could be used in the initial lift onto the Jebel Dhofar, most would be required to ferry ammunition, water and rations in the early days of the operation. The helicopters were needed to sustain the effort until an airstrip could be built and secured on the Jebel Dhofar, so that the helicopter effort could be replaced and expanded as Shorts Skyvan fixed-wing transports took over the supply and reinforcement effort.

Later that morning Omani aircraft brought in the other SAS squadron

The SAS and *firqats* from Mirbat and Sudh climbed the Jebel Dhofar from the east by a feature known as Eagle's Nest, and then worked their way westward during the day, thereby helping to divert the attentions of the *adoo*. Meanwhile, a full squadron undertook a gruelling march to occupy an old airstrip at a place called Lympne. This march still stands out in the minds of the men involved. The route was over very difficult terrain, and the bergen carried by each man contained enough ammunition and water to last for several days. But so severe was the march that upon their arrival even the SAS men were in no state to fight without a rest. Luckily the *adoo* were occupied elsewhere, and later that morning Omani helicopters and Skyvans started bringing in the other SAS squadron and the *Firqat Kalid bin Walid*. Colonel Watts decided to move to a more defensible position, so on the morning of the second day the SAS and *firqat* unit that had arrived by helicopter, – therefore fresher than the units that had marched in – set of for a location known as Jibjat. With the *Firqat Kalid bin Walid* in the lead, the Mountain Troop of G Squadron topped the small rise around Jibjat and came face to face with a large *adoo* group having breakfast. A firefight developed and the surprised *adoo* started to pull back in an effort to disperse, but not before a full-frontal attack was jointly carried out by the *firqat* and SAS, who overran the *adoo* position and continued to clear the area to the south, whereupon any further advance was stopped by a large wadi.

A heavy firefight continued that lasted until the SAS heavy support teams turned up. Each team consisted of a GPMG and three men: one carried the whole unit (gun plus tripod) while the other two carried ammunition and acted as gun loader and spotter (each man carried up to 1000 rounds of GPMG link ammunition – a ferociously heavy load). In the early days of the war these gun groups proved decisive in the winning of firefights, especially as the supporting Omani Strikemaster aircraft were at first not so quick in reaching the scene.

All through the second day of the offensive small battles could be heard flaring up at one location or the other. By the third day, Colonel Watts had split his force into three battle groups, two of which were dispatched to clear the Wadi Darbat and a ridge line known as the Gatn, pronounced 'Cuttin' by everyone.

The eastern battle group started to have problems with its *firqat*

For several days the *adoo* fought with everything they had, mistakenly thinking that this was nothing more than a short operation by the Omani forces and that in a few days they would give up and leave. It was not to be, though. The *Firqat Kalid bin Walid*, many of its men back in their own territory, fought as well as professional soldiers, bounding forward stride for stride with the SAS men. By 9 October the initiative was clearly on the side of the Omani forces, and the *adoo* broke up into smaller groups and disappeared into the small bush-covered wadis.

Meanwhile the eastern battle group, which had moved to a location called 'Pork Chop Hill', began to have problems with its *firqat*. This was not the first time the *Firqat A'asifat* had posed problems: from the start its men had been reluctant in many areas, and now they wished to observe Ramadan. So, despite the Sultan's urgings and the Qadi's dispensation, the *firqat* was withdrawn to Jibjat for the month of Ramadan. A few days later, Colonel

Watts descended on the *firqat*'s leaders and left them in no uncertain terms as to how he viewed their unprofessional attitude.

As both sides eased back a little, a main base was established at a place known as 'White City'. By this time the choppers were quickly running out of flying hours and desperately needed servicing. Likewise supplies of ammunition (especially mortar bombs) were dwindling, and water was at a premium. It was suggested by Colonel Watts that an airstrip should be constructed in the middle of White City so that the Skyvans could alleviate the supply situation. As troops began arriving at the location, the men of the *firqat* were sent to picket the high ground while SAS men set to work constructing the runway. They worked all night, several times coming under enemy fire, but by dawn the airstrip was ready to receive the first aircraft. Again the battle flared up. Each time a Skyvan landed the *adoo* were waiting: mortar bombs began to fall and small-arms fire was employed in an effort to shoot down several of the aircraft. To prevent this, heavily armed dawn patrols were sent out to engage and occupy the enemy while the aircraft unloaded. This period saw some of the heaviest fighting of the war.

It was an unbelievable sight when the first salvo landed

Despite the problems with the other *firqats*, the *Firqat Kalid bin Walid* in the western group continued to fight hard. When news arrived that a large contingent of *adoo* had been observed in the Wadi Darbat, most of them suffering from the terror of recent battles, the *Firqat Kalid bin Walid* set about planning a raid. Aerial bombing and artillery bombardment had little effect and ammunition was in short supply, but eventually a strong patrol was sent against the *adoo*-occupied village of Shahait. A fierce gun battle erupted which left two *adoo* dead, but many others escaped before the SAS and *firqat* could arrive.

From a distance of 5km (three miles), a large group of *adoo* was spotted by an SAS trooper who, pinpointing their position on his map, called in an artillery barrage from gun lines at Taqa. It was an unbelievable sight when the first spotting salvo landed smack in the middle of the *adoo*, and the trooper could be heard screaming into the radio: 'Fire for effect – fire for effect.'

In battle the *adoo* were courageous, always dashing into the fight

At first the conflict was fluid as the three main battle groups started in the east and pushed their way west. Those first weeks were by far the toughest and, as with all things new, it took time for the systems to fall into place and for effective coordination to be established with the supporting arms, such as the Omani artillery, Omani Air Force and, most of all, the *firqats*.

It would be remiss not to stress the importance of the bond between the *firqats* and the SAS. The *firqats* had certainly been trained to act as military units, but in reality they were far from being military units in any real sense of the phrase. Yet they possessed a feeling for their own back yard that the SAS did not have. It was not uncommon for their men to wander into battle with their rifles slung over their shoulders and then, quite suddenly, to drop to the ground and start darting forward. It was a movement the SAS soldiers came to recognise – it meant *adoo* were near. In battle they were courageous, always dashing into the fight even if sometimes their firing became a little erratic. They were also honest, and if for some reason the SAS unit did not do as they requested they would soon make their point obvious. At the same time, when they were around one was guaranteed a good night's sleep. Studying the *firqat* gave one some idea as to what the *adoo* were like (in fact, as mentioned above, many of the *adoo* who had been captured or surrendered themselves would help swell the ranks of the local *firqat*).

The *adoo* were mostly well equipped. It was not uncommon after the battle to find dead or wounded *adoo* dressed in a better kit than the SAS: khaki shorts and shirt, ammunition belt, water bottle and AK-47 assault rifle, all topped off with a blue beret complete with red star (plus a copy of the *Thoughts of Chairman Mao* in his pocket). Dead *adoo* were often stripped of their weapons and ammunition before being left by their comrades.

In the first weeks of battle contacts were very close and in large numbers. Rarely an hour passed without one of the three battle groups coming under fire. It was a matter of advance to contact, hold the firefight, bring up gun groups, and call in jets. Winning the firefight is the basis of all victories: hit the enemy with a wall of accurate fire and he will stop. For the SAS this firepower was supplied by two of the best infantry weapons ever: the 81mm (3.2in) mortar and the GPMG.

During a squadron patrol to clear a wadi of *adoo*, the commanding officer thought it prudent to protect his one exposed flank. The problem lay with a smaller wadi running parallel to the one the British and Omani force intended to search, for this offered the *adoo* excellent concealment and an escape route. To solve the problem it was decided that one SAS section, complete with about 10 *firqats*, would patrol the adjacent wadi. This worked out fine until the men reached a dry waterfall in the wadi bed. Luckily for the SAS party, its men happened to be at the top of the falls while a group of five *adoo* were coming up in the opposite direction. The *adoo* were trapped, isolated in what looked like a Roman amphitheatre. The *firqat* called out for them to lay down their arms as they were surrounded. A burst of automatic fire was the only reply, and

Below: Omani Skyvan transport aircraft prepare to land under fire at 'White City' during Operation 'Jaguar' in late 1971.

so battle was joined. For the SAS men it was a turkey shoot as they sat on the top of the falls and poured down fire. M79 grenade launchers, LAW rockets and full-automatic fire from at least 20 other weapons reduced most of the larger rocks in the area of the *adoo* to rubble. When at last the dust settled and a party went forward to investigate, nothing but a few bits of blood-stained cloth remained. Then it dawned on everyone that they had all missed and that the *adoo* had legged it. They gave chase, close enough for the blood on the rocks to be still wet. But the *adoo* on this occasion got away. There are many as yet untold stories of the Oman war, the most outstanding being the *adoo* attack at Mirbat (discussed below).

The war progressed into a steady round of stand-off raids by the *adoo*. In reply strong patrols and ambushes were carried out by Omani forces operating from their established

Left: Taqa. The SAS British Army Training Team (BATT) house is the building on the right with the white-framed door.

bases. But it soon became obvious that to stop the enemy completely would need two things: the severing of the *adoo* line of communication and re-supply, and then the winning of the hearts and minds of the peoples of the Jebel Dhofar itself. The first was a matter of laying minefields and establishing firm defence lines in order to monitor the *adoo* movements and respond accordingly. The second was a task at which SAS soldiers are masters.

To win over the locals small aid stations were set up and manned by SAS medics, while SAS Arabists would regularly talk to village leaders, and their problems became the problems of the SAS. Civil-aid teams soon moved into the liberated areas. Water was found and drilling teams brought it to the surface. In a land where water has priority next to life itself, the expression on a jebali's face is one of pure wonder and delight when the cool, clear liquid gushes from the ground. In addition, communications open up trade and commerce, and this became clear when at last a metalled road linked the Jebel Dhofar with Salalah.

One special operation known only to the SAS and a few others was Operation 'Taurus'. Thousands of goats and cattle were rounded up and purchased by the government for food, at the same time denying the *adoo* of one of its main sources of food. It was a truly remarkable sight as the animals were rounded up by the *firqat* and, amid fighting with the *adoo*, herded down the Jebel Dhofar to Salalah with Strikemaster jet support.

The bulky rucksacks contained mostly ammunition and water

With the establishment of the airstrips and a good re-supply system, more adventurous patrols were undertaken. One such task was the Jebel Aram. It was cold and dark as the patrol set off from White City at around 20:00 hours. The patrol consisted of some 15 SAS, 30 *firqat* and a platoon of Omani soldiers. The weight carried by each man made movement very slow. The bulky rucksacks contained mostly ammunition and water, but in addition three of the SAS men carried an 81mm (3.2in) mortar between them. This group made up the tail end of the lead group, and behind them each member of the Omani platoon carried two bombs. This was all the support the party would have until it had reached the Jebel Aram and established a strong base upon it.

Right: SAS troops on the Jebel Aram, October 1971, prior to the assault. Note the two GPMGs, which were used for support fire.

At last the men came to a halt, and with it the chance to rest from the merciless weight they carried. The SAS, apart from the mortar crew, and the *firqat* had gone forward for a reconnaissance. It was estimated that they were about 500m (1635ft) from their objective. The SAS and *firqat* had almost reached the old tree that was to serve as their objective (old trees in the desert are almost always used as landmarks). There was barely anything for them to see, and no obvious cover from where the enemy could spring an ambush. Thus they advanced fairly swiftly towards the objective.

Like all *firqat*, the men were sure-footed in their own back yard and presently they halted the patrol. Suddenly one of them dropped down on one knee and pointed, whispering '*adoo*' at the same time. The SAS used night scopes to survey the ground ahead. There, at the edge of their night vision, stood the ancient tree. The SAS commander signalled for the remainder of the group to close up and indicated the direction by pointing into the darkness. Then he quietly relayed the information back to the Omani commander. When he had finished he gave instructions to set up the mortar. As this was done, the men of the Omani platoon dropped off their mortar bombs at the mortar location, then spread out in defence covering the rear. Once everyone was ready, the SAS and *firqats* moved to the base of the tree.

From the silence and the darkness came a crashing orchestra of sound

The cold, damp atmosphere signalled that the coming day was not far away: even as they approached the tree, the sky to the east showed a thin, grey band. Then it happened. From the silence and the darkness came a crashing orchestra of light and sound, turning the desert into daylight and the weird tapestry of hell began. Everywhere men could be seen running or diving for whatever scant cover they could find. Almost instantly the SAS mortar thundered into action, the ground thumping as the bombs hit the pin at the base of the barrel and the propellant ignited.

'The base of the tree,' someone yelled, and blindly everyone fired in that general direction. Abruptly the steadfast sound of several 'gimpies' (GPMGs) chattered with reassuring effect. Suddenly the commanding officer's voice cut in with clear and concise fire orders. Slowly the firefight died, the silence interrupted only by long-distance sniper fire.

By daylight the patrol had settled down and set up a defensive position overlooking the Jebel Aram. This included the construction of a mortar pit and the clearing of a helipad for the delivery of ammunition, water and defence stores. That evening short patrols were sent down into the wadi situated at the back of the Jebel Aram. These small patrols could rely on accurate covering fire from the mortars and GPMGs, which all had spectacular views of the immediate area.

The author takes up the story: 'For a few days we settled in nicely, then it was decided to send out a day-time patrol down into the wadi bed. Things went well until they got "bumped" by the *adoo* and started to withdraw. The wadi bed was filled with short bushes and trees, making observation difficult for accurate mortar control with our own friendly forces in the area. So it was decided that a reinforced patrol would move farther down the hill to give support. It was a big mistake. The *adoo* had

Above: Members of the author's squadron shortly before they engaged *adoo* forces on the Jebel Aram in October 1971.

anticipated our move and had cleverly concealed themselves in the long grass. I was part of the reinforcements, and as we rushed forward they caught us in the open. The first thing I remember was seeing the *firqat* next to me stop in mid-air. His feet were off the ground and his rifle dropped from his hand. In that split second I watched a plume of bright red blood explode from his back – he fell dead. I saw the *adoo*. He was lying some 15m (50ft) away, behind a few rocks, his head and weapon the only part of him exposed. I thought of firing, but running downhill my momentum just seemed to pull me on. I saw the *adoo* fire and knew instantly that I was hit. Luckily for me, the burst hit the rocky ground just in front, before smashing up into my legs. I remember vaguely going up and turning over in the air, then I slammed back down onto the hard rock. Reality seemed a long way off as I collapsed. At first the pain was not so bad, then it registered and darkness came and blocked it out. Through dulled senses the noise of gunfire returned and I remember screaming for my partner, Jock Logan. Painfully I rolled my head to one side and saw the dead *firqat* close by. His chest was ripped wide open and bloody bubbles popped and oozed as air escaped from his shattered lungs. Suddenly Jock dropped down beside me. Quickly he started to pull the small rocks together to improve the protection around our shallow hole. Jock was breathless and panting like an excited dog. Finishing the meagre shield of rocks he thrust his "gimpy" over the top and sent a couple of blasts in the general direction of the *adoo*. Then he turned back to examine me.

"Its your knee, but it's not too serious." He was laughing as he applied the field dressing. "Medevac choppers are on the way."

'Some 20 minutes later I was on the operating table at the Field Surgical Hospital in Salalah. Jock had been right, the damage was light. Most of the bullets had hit the rocky ground first and splintered, the one which had stayed intact had lodged itself deep into my left kneecap. Whatever, two weeks later I was hobbling around the bars of Hereford. To this day I still have the AK-47 bullet head the doctors gave me.'

Despite the casualties, by the end of 1971 the SAS had established a presence on the Jebel Dhofar, and had saved the Sultan's regime.

Below: SAS mortar pit on the Jebel Dhofar after the successful conclusion of Operation 'Jaguar', which established an SAS presence on the jebel.

HOSTAGE-RESCUE

At Mogadishu in 1977 SAS soldiers assisted GSG 9 in a breathtaking hostage-rescue mission. Three years later the Regiment had its own success in the heart of the British capital.

Wadi Haddad's PFLP (Popular Front for the Liberation of Palestine) was going through rough times. His early hijack attempts had been successful, earning both him and his organisation worldwide notoriety. Then came Entebbe in July 1976. Haddad brought off the hijack of an Air France airliner using a joint West German Baader-Meinhof and Palestinian team. Despite the fact that the whole operation had been meticulously planned, and that the airliner had been hijacked to a country friendly to the PFLP (Idi Amin's Uganda), he had been humiliated. The Israelis fought back in a stunning fashion. In a superb military operation, a special anti-terrorist team flew thousands of kilometres in a daring rescue operation – and it worked. Entebbe remains to this day a milestone in hostage-rescue history.

But Haddad was not deterred. Despite his ailing health (he was dying of cancer), less than a year later he tried again. This time he hijacked a Lufthansa Boeing 737 which was returning from the holiday resort of Palma, Majorca. His team was all-Palestinian in the form of two men and two women. They had been trained in Aden, but they had flown to Palma from Baghdad. Although the hijack team

Left: Members of the German GSG 9 hostage-rescue unit practise building assaults. Hostage-rescue units must train vigorously to be successful in their endeavours.

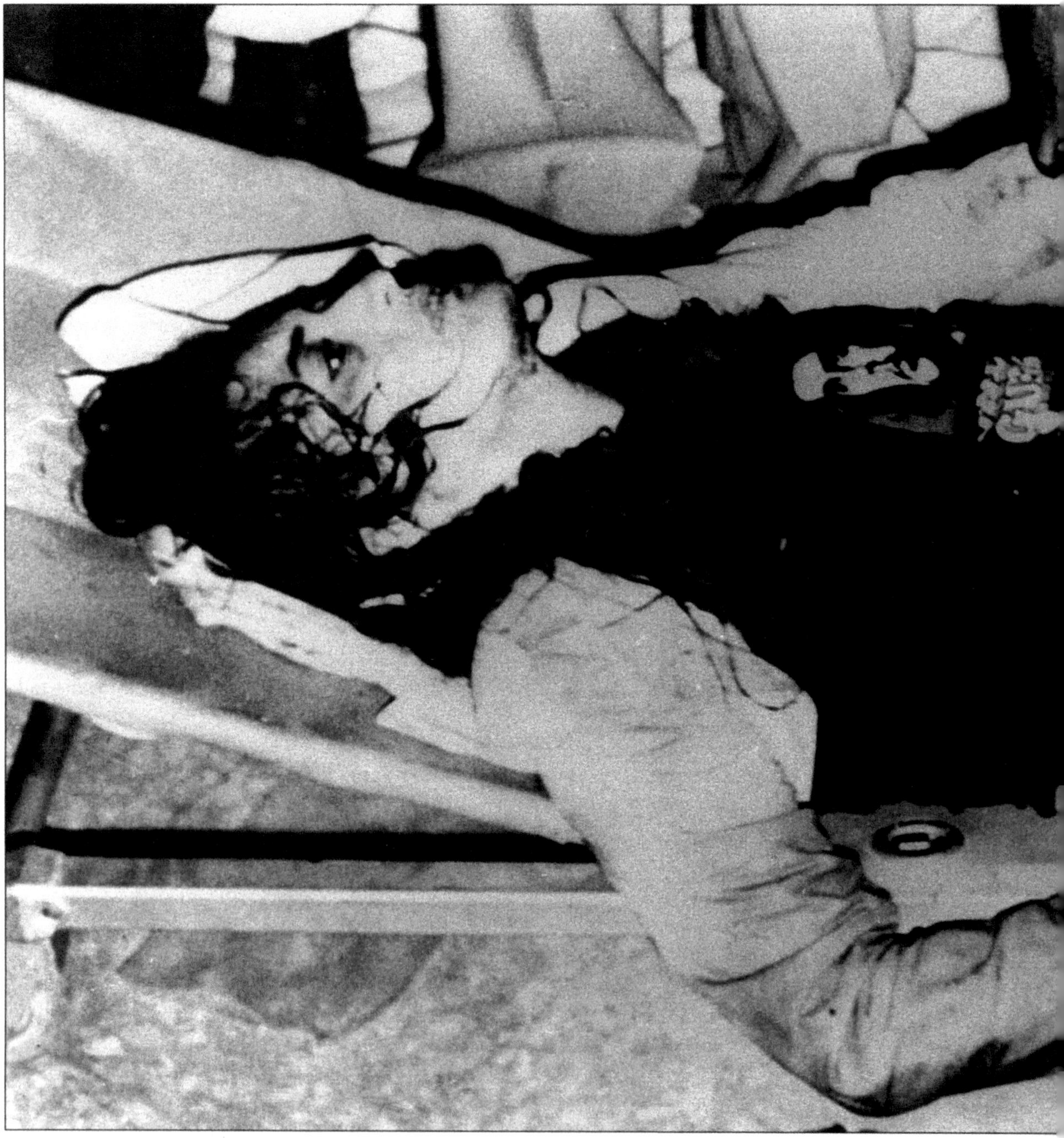

was Palestinian, the operation was jointly coordinated with the Baader-Meinhof group, which by this time had become known as the Red Army Faction (RAF).

For its part, the RAF had kidnapped a top German industrialist, Dr Hans Martin Schleyer, a month before. As it later transpired, both the kidnapping and the hijack were allied to the same demands.

The hijack team spent several days in Palma posing as tourists and selecting a flight that would take them to West Germany. After having visited several travel companies in the holiday resort, they eventually booked several

seats on Flight LH181 which was bound for Frankfurt. Later, it is reported, three of the group returned to the airport where they met a German woman who was pushing a child's pushchair. This woman handed over a biscuit tin to them which contained the weapons and grenades for the hijack.

Left: Soraya Ansari, the only member of the Palestinian hijackers to survive the Mogadishu rescue mission.

Everything seemed normal and relaxed as the LH181 flight took off from Palma for Frankfurt at 12:57 hours on the afternoon of Thursday 13 October 1977. Only the 11 German beauty queens attracted any particular attention among the 87 passengers. No one suspected that there were also four Palestinian terrorists on board. The journey should have been completed within two hours, but in fact it lasted five days – five days of hell.

The terrorists were now demanding the release of the Baader-Meinhof gang

The inflight meal had just been served when the hijackers struck. All four terrorists seized control of LH181, brandishing pistols and hand grenades, and starting a drama that was to ricochet around the Middle East. Refuelling at Rome and Larnaca in Cyprus, the airliner then flew on to Bahrain. Here it refuelled yet again before making its way to Dubai in the United Arab Emirates. The terrorists were now demanding the release of the Baader-Meinhof gang members being held in a top-security West German prison, as well as two other terrorists held in Turkish prisons. They also demanded US $15 million. The demand note was identical to that issued for the release of the kidnapped German industrialist Hans-Martin Schleyer a month earlier. During the first 48 hours of negotiation, the Bonn government took a firm line, refusing any concessions – holding to a strategy of continuous dialogue as a means of achieving the hostages' safe release. At the same time, the West Germans looked to Europe for support in its stand, and received assurances from both France and the UK.

The author, who with Major Alastair Morrison assisted the German anti-terrorist team throughout the hijack, takes up the story: 'During the week preceding the hijack, I was on duty at Heathrow with eight other members of the SAS anti-terrorist team. We were training on various types of aircraft, familiarising

ourselves with their basic internal layouts and the many variations employed by different airlines. Most of our work was done during the periods allowed for cleaning between scheduled flights. On the afternoon of Friday 14 October, I returned with my crew to Hereford through the beginnings of fog which was thickening dramatically across the country. On arrival at Stirling Lines we checked in all our equipment and our vehicles. Ensuring that all my men were on call, I released them for the weekend before setting off for my own home. I had only just arrived when the phone rang, ordering me back to camp. Back at the anti-terrorist team office I found the team commander, Captain Holmes, who informed me that the British and West

Left: A GSG9 operative uses a ball mount to the windscreen to fire his MP5 SMG. It was after Mogadishu that the SAS adopted the MP5.

German governments had agreed on the need for a joint anti-terrorist effort, and that the two of us were to leave immediately for London. As the fog was so thick we took a chopper, which flew us directly to Battersea Heliport.

'Unfortunately the heliport had been closed by the fog, so that when we landed we had no option but to climb over the heliport gates to get into the street. Luckily, one of the first vehicles we saw was a police panda car, which we flagged down. We asked the driver to take us to Whitehall. He was, not surprisingly, dubious about such a request from two scruffy individuals, but our manner, backed up by ID cards, persuaded him to check with his control. They played it safe, telling the PC to give us a lift to see if we were genuine – and to bring us in if we weren't!

'When we arrived outside No 10, we were met by senior military personnel, who briefed us on the current situation. Major Morrison of SAS HQ in London joined us at this time. We were further briefed regarding the areas of national secrecy concerning the operation and the equipment employed. It was made clear that we were to give all possible assistance to achieve the release of the hostages, but were instructed not to talk about some of the recent equipment advances we had developed.

'We attended a meeting inside No 10, at which the situation was fully discussed'

'We got the impression that a couple of politicians from Bonn had arrived, together with two members of a unit barely known to us at the time – GSG 9 (*Grenzschützgruppe* 9, a division of the boarder police trained in anti-terrorism). We attended a meeting inside No 10, at which the hijack situation was more fully discussed. Present were various heads of security departments, ministers, Major Morrison, Captain Holmes and myself. It was quickly established that the plane's position was still as reported by the media that afternoon: in Dubai. We in turn reported that there was an ex-SAS man currently in position in Dubai working under contract for the Palace Guard.

'We were then introduced to the GSG 9 members, and within minutes realised how much in common the SAS Anti-Terrorist Team had with them, for we had each developed tactics and equipment which would later be of great benefit to both units. I mentioned that we had developed a new type of stun grenade which would detonate almost instantaneously when thrown, effectively stunning anyone in close proximity. The grenade emits a very loud bang and a very bright flash of light in a set sequence – not dissimilar to the effect of strobes in a disco as they flash on and off. The effectiveness of these grenades, together with our expertise and knowledge gathered from the Middle East, and our recent training on aircraft interiors, was of such value to the Germans in dealing with the hijack that we were immediately asked to return with them to Germany. They also suggested that once we had talked to the people in Germany, we could, if requested, fly on to Dubai to give further assistance. Major Morrison and I were selected to accompany GSG 9, and arrangements were made for eight stun grenades to be sent from Hereford to meet us at Brize Norton.

'The language was pretty blue as some shocked GSG 9 soldiers emerged'

'We left shortly afterwards, the intention being to travel by helicopter. However, due to the very thick fog we had to endure a very tedious drive – sometimes at not more than 16-25 kmh [10-15 mph] – arriving at Brize Norton at about 04:00 hours. Here we met the crew of the C-130 which had been placed at our disposal and was ready for an immediate take-off. The weather was so bad we crossed the channel using low-level radar during the flight.

'Already aboard – and under guard – were the two boxes containing the stun grenades. I checked them at once to make certain that we had fully operational grenades, and not the training variety. The C-130 then took off, landing in Bonn, West Germany at 06:30 hours on Saturday 15 October.

'There to meet us were the two GSG 9 Officers we had previously met in London. They immediately took us directly to GSG 9 HQ. There we had a very short discussion with the unit's second-in-command (the commander, Ulrich Wegener, was already in Dubai), and it was decided we should demonstrate the British stun grenades to the Germans. The most convenient space, which was similar in size and shape to the interior of a plane, was a long corridor in the cellars of the HQ building. About a dozen GSG 9 soldiers took up positions in various recessed doorways. With

Right: A GSG 9 combat team landing during a training exercise. The unit is closely affiliated to the SAS.

the lights out I tossed in a stun grenade. The language was pretty blue as some very shocked GSG 9 soldiers emerged from the cellar corridor, but nevertheless it proved how effective the grenades were. The second-in-command then made an instant decision to send both Alastair and myself on to Dubai by the speediest means. Unfortunately, this meant getting the 12:12 hours plane out of Frankfurt for Dubai, and then changing in Kuwait. All went smoothly until we arrived in Kuwait. As one can imagine, the entire Middle East was alarmed by the hijack in Dubai. For this reason Kuwait Airport was on full military alert, and even passengers in transit had their luggage re-checked before being allowed back aboard the plane for Dubai.

'Although Alastair and I hung farther and farther back in the queue, it was inevitable that we had to put our bag containing the boxes of stun grenades through the X-ray machine. I can

still envisage the screen clearly showing those grenades, and I still recall, very vividly, the commotion it caused. We were at once slapped under heavy guard and manhandled quite ruthlessly by Kuwaiti soldiers into the main Security Officer's room, followed by our suspicious hand luggage.

'The bag was opened for examination, a procedure which I had to terminate when one of the Arabs tried to remove the pin from a grenade. At this stage, realisation dawned on everybody in the room that we were taking the grenades from Germany to Dubai. Luckily for us, at that moment the general manager for Lufthansa in Kuwait came into the security room. After a few moments he left the Security Officer in no doubt that, unless we were released immediately, together with our

Left: Germany's GSG 9, along with the SAS, is one of the most prominent members of the anti-terrorist brotherhood.

grenades, to rejoin our original flight to Dubai, no other Lufthansa aircraft would ever fly again into Kuwait Airport. It worked, and we were hustled across the tarmac to the waiting plane before physically being pushed into our seats. Then the bag with the stun grenades was dropped into my lap for me to nurse.

'We arrived in Dubai at around 03:00 hours on Sunday the 16th – only to be arrested at the airport for lack of documentation. Our passports were taken from us, and this action allowed one of the Western news reporters covering the hijack to pick up our names. Later he realised our true identities, and that we were SAS. Alastair made numerous attempts during the early hours of the morning to contact the British Embassy, but got no further than the gate man who was manning the night phone. Then luck came our way in the form of an ex-SAS officer called David Bullig. He had left the SAS and had been seconded to work training the Dubai Palace Guard. When he saw us both he knew immediately who we were, and events took a completely different turn. Within minutes we were able to roam freely about the airport to assess the situation. David was extremely helpful in many other ways, not least because he had already primed some of the best men of the Palace Guard to be ready to attempt an immediate assault on the aircraft should the terrorists actually start killing the hostages.

All three of us left for the airport to meet GSG 9 and other British officers

'We toured the airport and then spoke to Wischnewski, the German minister who was the acting representative for the West German government. We also met and talked to the defence minister of Dubai, who had taken charge of the situation directly (this was the second hijack he had dealt with). Having fully assessed the situation, we went with David Bullig to talk directly to Colonel Ulrich Wegener, head of GSG 9, who was resting with several other Germans in the airport hotel. We all agreed that there was very little we could do until the morning. We would be better off refreshing ourselves with a little sleep and meeting later that day.

'David Bullig took Alastair and myself to his home, where his charming wife plied us with sandwiches and coffee while we laid out our plan of action. David scribbled notes, listing our demands for kit and equipment. Our most expensive request was for the use of a Boeing 737 for training and practice purposes. At around 05:00 hours both Alastair and I fell asleep. After a couple of hours' dozing we were awakened by Mrs Bullig with the news that David had managed to fulfil most of our demands. All three of us left for the airport to meet GSG 9 and three other British officers David had found from various units in Dubai. In addition, he had selected eight of the best men from the Dubai Palace Guard.

'We set out a very quick immediate action drill to meet the needs of the worst possible scenario. That's when we would have had no choice but to assault the aircraft with the limited force available. Of course, the more time we had the more our plan would improve. By now most of the kit and equipment we needed had arrived: shotguns, masking tape, walkie-talkies, various ladders, padding and a host of other items. We had two quartermasters standing by with four jeeps and an apparently endless supply of cash to obtain anything else required. Most importantly, a Gulf Air Boeing 737 had been given on loan and was parked at the far end of the airfield, well out of sight of the hijacked plane. I was about to start a crash course in anti-terrorist techniques.

'The Boeing 737 is a simple animal where anti-terrorist drills are concerned'

'On the personnel side, my resources were limited to a hard core of five men who had received at least some professional Close Quarter Battle (CQB) or anti-terrorist training. These included Alastair, David, myself and the two GSG 9 officers. Additionally, I had three other British officers and the eight soldiers from the Palace Guard. I concentrated our first efforts on the immediate action drill needed to counter any terrorist deadline.

'The Boeing 737 is a simple little animal where anti-terrorist drills are concerned. There are only three options for entry: tail, wing and front catering area. We thought that if the terrorists began to carry out any threatened shootings, they would naturally take the precaution of covering the main doors. It seemed less likely that they would cover the two emergency exit doors leading onto the wings, so the basic plan that fell into shape was to attack through these entry points. The fact that the wing emergency exits were designed to be opened easily from the outside was another strong factor in favour of adopting this mode of attack. In addition, we had also discovered a blind spot were the wing joins the aircraft body. Two men could easily sit beneath the emergency doors and not be visible from any of the plane's windows.

Left: The Regiment has integrated all the lessons learnt at Mogadishu into its aircraft hostage-rescue training.

'The leading teams were to receive back-up from the second assault pairs'

By comparison, the entry and exit points at both the front and rear require considerable manhandling and some time to get them open. For instance, the front door is operable through a small hatch on the outside of the plane which allows the door to be opened and brought down and the stairs to extend automatically.

'The basic moves involved in our plan were:

'First, to make a single-file approach to the airliner from its blind spot at the rear, assemble our ladders quietly and erect them to the wings and the rear door.

'Second, to put each of the two leading assault teams covertly on the wings – one outside the port emergency exit, the other outside the starboard – with the other assault teams waiting on the top rungs of the ladders. Each of these assault teams consisted of one SAS or GSG 9 man backed up by one of the best of the soldiers from the Palace Guard. The reason that we involved the Palace Guard in this way was purely political (the Dubai government being unwilling to allow us to mount the assault unless it played a part).

'Third, to position back-up squads beneath the rear area of the airliner ready to scale the ladders, open the rear door and effect their entry as quickly as possible. At the same time, a

EXIT

Above: The SAS offered assistance during the Dutch De Punt train incident where South Moluccan terrorists held up a trainload of people.

second back-up squad would move quietly forward until they were beneath the front door area, ready to erect their ladders and follow suit. The back-ups would coordinate their moves with those of the assault teams.

'Fourth, when everyone was in position and the "GO" command had been given, the leading assault teams were to stand, punch the emergency exit panels and drop the doors into the laps of the passengers in the mid-section of the cabin. The teams would then enter, the port pair clearing to the front of the cabin, the starboard team clearing to the rear. The leading teams were to receive immediate back-up from the second assault pairs entering behind them from their stations at the tops of the ladders to maintain control of the centre of the airliner.

'Fifth, simultaneously with the assault the outside squads were to open both front and rear doors and enter the airliner. Their intention was to provide further back-up in case of any problems, and also to provide routes for the hostages to leave the plane, which by this time would be full of smoke from the stun grenades.

'As I mentioned earlier, the Boeing 737 is a fairly simple animal. Once entry had been made at the centre of the airliner, the starboard assault team would have a clear line of sight to the toilet doors at the rear of the cabin. The port team, moving forward through the economy area, would arrive in the first-class section which leads into the front catering area. Directly beyond this is the flight deck, the door to which is usually closed. The only obstacles that the team would encounter would be this door and at the curtains that separate the economy and the first-class sections.

'Although this basic plan was quite uncomplicated, we calculated that it would require a great deal of practice to get the timing right, especially the time it would take the assault teams to effect their entry and make their way to the front and rear of the passenger cabin. We reckoned that as soon as we dominated these points, the only people in serious danger would be the crew in the cockpit.

'We refined our assault plan as best we could and developed external additions'

'By 08:00 hours on the 16th – 67 hours after LH181 had left Palma – the training and practice began. We took a break around 14:00 hours, using the time to iron out every detail, searching for anything that might increase the odds in our favour.

'One thing that did improve the odds were the Palace Guards. I was truly amazed at the quality of the soldiers we had with us. They turned out to be extremely tough, very quick and could be relied upon to do exactly as they were told in what was, for them, an entirely new kind of situation. We now felt fairly confident that if the terrorists forced us into immediate action we had a better than average chance of success. We were now sure that we could approach the airliner and establish all our people in their starting positions unobserved. We knew we could put eight men – the four assault pairs – inside the airliner. We knew that the outside assault teams were fully aware of their duties regarding getting the doors open, effecting their own entry and giving assistance where necessary.

'We refined our assault plan as best we could and then started to develop external additions. We worked out a suitable distraction for the terrorists. We considered how best we could locate their positions inside the airliner – extremely valuable information if we could get it. We also gave considerable thought to the possibility of an attack at night, during which, if it came off, we would have an extra man underneath the plane to shut down the APU (Auxiliary Power Unit). This is located at the rear of the wheel housing, and one yank on the red and yellow handle would kill all the lights and power in the airliner at the moment which suited us.

'At about 15:30 hours the Dubai defence minister left his watch in the control tower and came to check on our progress. We went through our operation on the Gulf Air Boeing 737, and I have to admit that it looked pretty impressive. Then, just as we finished our demonstration, the unexpected happened. LH181's engines came to life and took off, taking with her any hope that the hijack would end in Dubai.

'Fortunately, the Germans had a Boeing 707 for the use of the negotiators. Everybody from Germany concerned with the hijack, as well as Alastair and myself, went aboard and we took off to follow the hijacked plane. First indications were that it was going to land at Salalah in southern Oman. This sounded like good news, for the SAS had men in the area with anti-hijack skills. Our expectations were dashed, however, when we learned that LH181 had in fact landed at Khormaksar Airport in Aden. Captain Schumann, piloting LH181, had been too low on fuel to fly anywhere else, and he skilfully put the aircraft down on the hard sand alongside the runway.

'Wherever LH181 was forced to fly, we would make our rescue attempt'

'Our Boeing 707 had to fly on to the international airport in Saudi Arabia. Here, confined in the aircraft, we sat on the ground awaiting further developments. During this waiting period the negotiating psychologists attempted to persuade the hijackers to release the hostages in exchange for the $15 million, which we held in a large suitcase on board our

Right: London, 5 May 1980: Members of Red Team on the roof of the Iranian Embassy prior to their assault.

plane. Then came the shocking news which brought an immediate end to negotiations.

'Captain Schumann had been shot dead aboard LH181. Now the decision more or less made itself. Wherever LH181 was, wherever she was forced to fly, we would make our rescue attempt. For the first time I saw the true determination of our German partners concerning their own nationals. The killing of Captain Schumann decided the matter. Any further ideas about peaceful negotiation were instantly dismissed.

'Our next news was that LH181 had again taken off from Aden with the dead pilot aboard, and had flown to Mogadishu, capital of the Somali Republic. We got airborne in the Boeing 707 and sought permission to land there as well. Our flight from Saudi Arabia to Mogadishu took us directly across the war zone lying between Ethiopia and Somalia.

'The terrorists announced that their deadline was 15:00 hours'

'On approach to Mogadishu we were given permission to land, but the situation was complicated by the presence of LH181 sitting in the middle of the main runway. Our pilot was equal to the challenge, however. Using only a short length of runway, he brought off a superb landing, using every metre of concrete available and rolling to within literally metres of buildings and houses on the airfield perimeter. Even the intensity of the overall situation could not obscure this brilliant feat of professional skill, and everyone aboard broke into spontaneous and grateful applause.

'On the ground two top Somali officials collected all passports – and were surprised to discover two of the British variety among them. The officials were most courteous and friendly. The German minister went off to meet the Somali prime minister to discuss the developing political aspects of the hijack. The rest of us were taken to one of Mogadishu's top hotels and given accommodation and a meal. We were kept in constant contact regarding the hijack situation through the Mogadishu security services, who were very friendly and helpful.

'The terrorists now announced that their deadline was 15:00 hours on that same day: Monday the 17th. Negotiators from Mogadishu tower asked for an extension, explaining that

the Baader-Meinhof terrorists jailed in Germany, together with the two Palestinians held in Turkey, would be released. They could not, however, be flown to Mogadishu in less than 10 hours. After prolonged discussion the terrorist leader, 'Captain Mahmoud' (Zohair Akache), agreed on a final deadline of 03:00 hours the next morning, Tuesday 18 October.

'In the meantime, Minister Wischnewski obtained permission to bring in the German anti-terrorist team, flying to Mogadishu with the option of an assault on LH181. While awaiting its arrival, we all worked together to modify the assault plan to match it to the current situation. Joining us in our planning was a colonel from the Somali Special Forces.

We stuck to the basic plan to approach the aircraft in single column from the rear, breaking into four sub-sections, each moving to its assigned position with the intention of making simultaneous entries through both wing emergency exits, the port front and starboard rear doors. We modified the choice of ladders for the assault on the doors, using double (side-by-side) instead of single ladders. The advantage of these was that two men could go up together. At the top, the left-hand man could turn the door handle and swing his full weight away from the fuselage, so pulling the door open quickly. This made it possible for the second, right-hand man to penetrate the airliner immediately.

'I was also to throw a stun grenade over the front of the cockpit'

'We needed outside help for two reasons. First, before the assault began we wanted the negotiators to start talking positively to the terrorists to ascertain as closely as possible their whereabouts inside LH181. We also thought we might encourage them to come to the cockpit by lighting a very large fire at the far end of the runway. This job was entrusted to the Somali soldiers, who were also responsible for ground defence around the aircraft.

'The GSG 9 men arrived in a second Boeing 707 at 20:05 hours on the evening of 17 October. Their commander briefed them immediately, and they set about preparing their equipment for the assault. We ran through a quick rehearsal, using the GSG 9 Boeing 707, which was parked out of sight of LH181. By 23:30 hours the whole group was in position about 70m (225 feet) to the rear of the airliner. In single file we approached the plane, and in complete silence the ladders were put in position at each wing root and against the chosen doors. Major Morrison and I were on either side of the fuselage, at the rear of the wing roots. Our initial task was to throw stun grenades over the fuselage just as the doors were opened, to achieve the penetrating effect of noise and light in the cabin. I was also to throw a stun grenade over the front of the cockpit to assist further in the disorientation of the terrorist leader.

'The approach to the aircraft was very slick and smooth. The only problem we had was that the airfield lighting around the control tower created long shadows. Had the window blinds been up, any one of the terrorists looking out could have seen us approach. But this was not the case. The GSG 9 commander was in direct contact with the tower. Just before the operation began he transmitted to all his assault teams that the two male terrorists had been heard in the cockpit. At this moment the fire was ignited at the end of the runway. In spite of the tension, this caused some amusement, for it was plain that the Somali soldiers had let their enthusiasm run away with them. It looked as if they had set fire to a complete tanker-load of petrol.

'The GSG 9 commander counted down and gave the "GO" signal. Everything happened at once. The quiet African night erupted. The front and rear doors opened as the left-hand ladder-men swung on them; their right-hand partners heaved themselves out of sight into the plane. The wing assault teams stood up, punched the emergency exit panels and vaulted in as the doors fell into the cabin – all these actions orchestrated by the bangs and flashes from the stun grenades. Immediately the rear starboard door swung open the first terrorist was sighted, absolutely amazed by this turn of events. She was shot instantly by a GSG 9 soldier, who then threw himself flat into the rear catering area alongside the toilets, firing up the aisle where the other female terrorist had been spotted.

'During the firefight we heard two dull explosions inside LH181'

The front assault team was involved in a brief firefight with the two male terrorists. Lasting for about a minute, it ended when they had both been fatally shot. During the firefight we heard two dull explosions inside LH181. These were hand grenades exploding as the terrorists' strength drained away and their grip on the grenades they were holding relaxed involuntarily. Thus, as the first few minutes of Tuesday 18 October ticked away, the hijacking of Flight LH181 ended.

'As soon as the firing died away the passengers began to disembark. And now occurred one of those unforeseen circumstances which could have led to quite serious injury among the passengers, who seemed convinced that every airliner is equipped with those inflatable rubber chutes into which they can jump and slide happily down to the ground. In this case they didn't exist, but several of the passengers, young and old alike, tried to slide down the assault ladders. To say that this caused us some concern would be an understatement. The thought of the passengers injuring themselves this late in the game was too much to take. The GSG 9 men on the ground took swift and firm command of the exits, guiding people down the assault ladders or assisting them through the mid-section emergency exits down to the ground via the wings. The passengers were in a state of sheer bewilderment. They had, after all, spent five long days cooped up in a very confined space, in hot, filthy conditions and failing sanitary facilities. On top of all this physical discomfort, every hour would have been heavily weighted by the fear feeding on the uncertainty of their future. The climactic few minutes of the assault, involving loud explosions, flashes, smoke, rapid movement, gunfire and raw danger, must have disoriented many of them. As they disembarked they were ushered to the rear, where a fleet of ambulances and other vehicles ferried them to the passenger lounge in the terminal building.

Below: The SAS assault goes in as flames, billowing smoke and CS gas fills the Iranian Embassy's rooms.

'The casualty list showed three of the terrorists dead – two men and one woman – and the second woman terrorist severely wounded. Incidentally, as she was taken away for medical treatment she gave the V-sign and screamed an assortment of slogans. One member of GSG 9, one member of the air crew and five of the passengers were slightly wounded. After any necessary medical attention, the passengers were soon taken aboard the negotiators' Boeing 707, while the negotiators, the GSG 9 men and we two SAS men boarded the GSG 9 707 to be flown back to Germany.

'Although apparently over, the hijack of LH181 still had a twist in its tail. There were still questions – unanswered questions to which no answers have been found. During the flight we heard on the radio that the leading members of the Baader-Meinhof terrorist gang, confined in separate cells in Stammheim Jail in Stuttgart, had committed suicide. Andreas Baader and Jan-Carle Raspe had shot themselves; Gudrun Ensslin hanged herself; while Irmgard Muller had made an unsuccessful attempt to kill herself using a stolen bread knife. How did the terrorists, locked in separate cells in a maximum-security prison, simultaneously learn of the German government's success in Mogadishu – 5600km (3500 miles) away – within a very few hours of its occurrence? How did the two male terrorists obtain pistols and ammunition in that same jail? These mysteries remain unexplained to this day.

'How did two male prisoners obtain pistols and ammunition?'

'Upon arrival at Frankfurt we SAS men were separated from the GSG 9 people and taken to the VIP lounge. Here we were instructed to retain our scruffy appearance until after our flight to the UK. Arriving in London at around 21:00 hours we were moved off to a secret rendezvous for a full debriefing regarding the operation. So came the final end to the hijacking of LH181, together with the SAS/GSG9 operation to rescue the hostages and crew. The success of our joint operation was marred, for me, only by the death of Captain Jurgen Schumann, shot in Aden. He was a very brave man and, like so many other aircraft captains, upheld his professional responsibilities to the end by seeking to protect all those in his care, passengers and crew alike. The success was soon worldwide news, and governments all over the western world applauded West Germany's determination in the face of an international terrorist threat.

'As a postscript to this account, it may be worth mentioning the kidnapping of Dr Hans Martin Schleyer, which took place just before the hijack. There were suspicions that the two events were linked in some way. Within a few days of the rescue, Dr Schleyer's body was found in a car boot with three bullets in the back of his head.'

It all started when a group of terrorists arrived in London

(Author's Note: The German woman reputed to have delivered the weapons at Palma is claimed to be Monika Haas, a one time member of the Baader-Meinhof gang. Haas was arrested for offences related to the Mogadishu affair several years later. In the spring of 1992, after German unification, Federal lawyers had come across remarks in documents of Section 22 (anti-terrorism section) of the former East Berlin Ministry of State Security (Stasi), according to which Haas is supposed to have delivered weapons to the hijackers. However, after seven weeks of interrogation, the Federal Court of Justice dismissed the arrest warrant.

The new arrest of Haas came after German investigators apparently located a Palestinian woman, Soraya Ansari, the only member of the hijack group who survived the GSG 9 mission at Mogadishu. On 13 October 1994, 17 years to the day after the hijacking, she was arrested in Oslo. During a 48-hour, non-stop interrogation, German BKA (Federal Criminal Office) officials questioned Ansari, who acknowledged her participation in the seizure of the aircraft. It is assumed that the two women probably know each other from the Yemen. Whether or not they also met on the holiday

Above: The outside of the SAS's 'Killing House' at Stirling Lines, Hereford, where hostage-rescue drills are taught.

island of Palma when the weapons were delivered is a question the Karlsruhe attorney-general (GBA – *Generalbundesanwalt* – or prosecutor's office) has been trying to establish for three years. While they do, both women find themselves prisoner: the Palestinian Ansari, 41, in temporary extradition custody at her home in Oslo, where she has lived since 1991, and the German Haas, 46, in a German jail. The authorities believe that both women belonged to the Palestinian PFLP, which had its most important base in a desert camp barely two hours' drive from Aden, the capital of The Yemen Republic.

Like the cadres of other terrorist organisations, the officials of the PFLP also travelled in and out of the former Eastern bloc, coming to the attention of Stasi agents. The PFLP member Zaki Helou, for example, was a regular customer of the notorious KINTEX weapons factory in Sofia, which was controlled by the Bulgarian secret police. Helou was formerly married to Monika Haas, but despite this close connection Haas denies taking part in any terrorist activity.

In January 1995 the author met Souhaila in Norway and tried to save her from extradition. He failed and she was sent to Germany to face trial in December 1995. The reasons for this strange action can be found in a true account that was recently published by Barry Davies, *Shadow of the Dove*.)

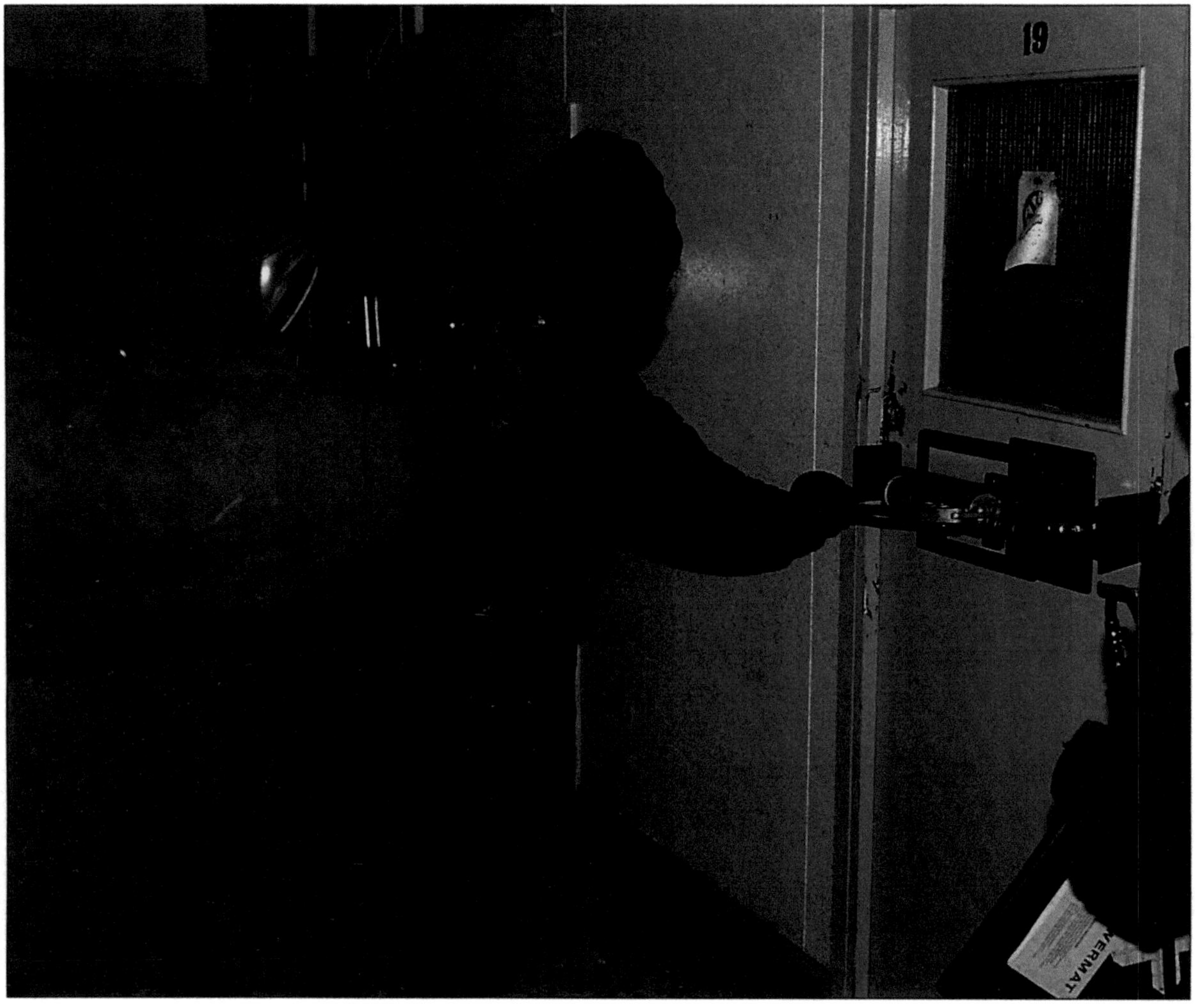

Above: Inside the 'Killing House', entry drills are practised over and over again until they become second nature.

The SAS had its day in May 1980, although by this time the anti-terrorist team had plenty of experience. Members of the Regiment had been present when the Dutch Marines had stormed into the train that had been taken over by South Moluccan terrorists (June 1977), and had assisted the West German GSG9 throughout the Mogadishu hijacking (October 1977). When their own time came, the SAS assault on the Iranian Embassy in London turned out to be a real classic. Moreover it was filmed on national television and beamed around the world.

It all started when a group of terrorists arrived in London, seeking accommodation. Almost inevitably they finished up in the Earls Court area, home to the influx of many foreigners. However, like so many members of the visiting Arab community, the pubs and clubs of the district are a temptation too good to miss. This mixture of drinking and womanising increased to the point where the terrorists were forced to leave their accommodation and seek fresh lodgings. Eventually, the whole group finished up in 105 Lexham Gardens, where they remained until their departure on the morning of Wednesday 30 April. By 11:20 hours, having collected several weapons, two of which were machine guns and a quantity of grenades, the group of six men stood outside No 16 Princes Gate, a smart, tree-lined avenue situated in London's Kensington district.

At 11:25 hours the six armed terrorists took over No 16, which housed the Iranian Embassy. The terrorists were opposed to the regime of Ayatollah Khomeini and were seeking liberation of Khuzestan (a region of Iran peopled not by Iranians but by ethnic Arabs) from Iran. As they took control of the embassy, they seized 26 hostages, including the British policeman who had been on duty at the entrance. This was something that might have gone unnoticed but for the fact that minutes later a burst of machine-gun fire could be heard. The police were on the scene immediately, swiftness initiated by the captured policeman, Trevor Lock, who had managed to alert his headquarters before being taken by the terrorists. Armed D11 marksmen soon surrounded the building and negotiating plans were immediately put into operation.

An intelligence cell was set up to gather and collate every snippet of information

The anti-terrorist team at Hereford was at this time practising in the 'Killing House', but things were soon to change. By 11:47 hours, Dusty Grey, an ex-D Squadron man who now worked with the Metropolitan Police, was talking to the Regiment's commanding officer at Stirling Lines. His information contained only the briefest details, but was enough to alert the Regiment. Several minutes later, the duty signaller activated the 'call-in' bleepers carried by every member of the anti-terrorist team. Although the SAS had prior warning, there can be no move before official sanction comes from the Home Secretary, who at the request of the police will contact the Ministry of Defence. Despite this red tape, it makes sense for the SAS to think ahead, and time can be saved by positioning the anti-terrorist team closer to the scene. Around midnight on the first day, most of the team had made their way to Regent's Park barracks, which had been selected as a holding area. From there various types of information could be assembled and assessed. The construction of a scale model of No 16 Princess Gate was ordered, and this task fell to two pioneers drafted in from the nearby Guards unit. Additionally, an intelligence cell was set up to gather and collate every snippet of information that would aid any assault.

By this time the terrorist leader, Oan, had secured his 26 hostages and had issued his demands. These included the autonomy and recognition of the Arabistan people, and the release of 91 Arabistani prisoners who were being held in Iranian jails. The terrorists threatened to blow up the embassy and kill the hostages, but by Thursday 1 May they had done nothing but release a sick woman. Later that day Oan had managed to get a telephone call through to Sadegh Ghotzbadeh, Iran's foreign minister. The conversation did not go well, the minister accusing Oan of being an American agent and averring that the Iranian hostages held in the embassy would consider it an honour to die for their country and the Iranian revolutionary movement.

This lack of cooperation almost forced the terrorists to seek a more sympathetic mediator, but it also meant that the Iranian government did not give a damn if their embassy staff were killed or not. Another problem had also raised its head for the terrorists. Chris Cramer, a BBC sound man, had become sick with acute stomach pains. His partner, BBC sound recordist Sim Harris, pleaded with Oan to call for a doctor immediately. This was done, though the police refused to comply straightaway. However, in the end Cramer was released and stumbled out of the embassy door and into a waiting ambulance.

If the terrorists started shooting, the SAS men would batter their way in

Later that night, under cover of darkness, three Avis rental vans pulled up in a small side street by Princes Gate. Men carrying holdalls quickly made their way into No 14, just two doors down from the site of the siege. Within minutes they had laid out their equipment and made ready for an Immediate Action (IA). At first the plan for this IA was very simple: if the terrorists started shooting, the SAS men would run to the front door of No 16 and batter their way in. The plan was brutal and primitive, but better than nothing until a more clearly and accurately defined plan could be organised.

By 06:00 hours on the morning of Saturday 2 May, the situation inside No 16 was getting very agitated. Oan used the specially established telephone link between the embassy and No 25 Princes Gate (The Royal School of Needlework) that now housed Alpha Control and the police negotiator. Oan's main criticism was that there had been no media coverage of the siege, and therefore no opportunity for his cause to be explained. After venting his frustration, Oan slammed down the phone in rage. By the late afternoon on the same day, Oan was allowed to make a statement, which was to be broadcast on the next news slot in return for the release of two more hostages, including one pregnant woman. The trouble lay in the fact that Oan would not release the hostages before the statement was read out. However, the police wanted the hostages to be released before any statement was broadcast. In the end a compromise was reached, and the broadcast went out on the mid-evening TV news.

Various sound distractions were supplied by men working on gas supplies

Two hours later eight members of the SAS team had climbed onto the rear roof of No 14 and were making their way through a jungle of TV antennae to No 16. Two of the men made their way directly to a glass skylight, and after some effort they managed to get it free. The skylight opened directly into a small bathroom on the top floor of the Iranian Embassy, and would provide an excellent entry point. Meanwhile, other members secured abseil ropes to the several chimneys and made ready for a quick external descent to the lower floors, where they could effect an entry by smashing through the windows. Oddly enough, an enterprising TV director had managed to get a camera into a bedroom window overlooking the back of the embassy, and during the assault he filmed the whole action.

Right: Deep inside the 'Killing House', a fully equipped trooper aims at cardboard targets with his Heckler & Koch MP5 SMG.

By 09:00 hours on Sunday morning things appeared to be heading for a peaceful settlement. Oan had agreed to reduce his demands, and at the same time some Arab ambassadors had attended in Whitehall a Cabinet Office Briefing Room (COBRA) meeting that was chaired by William Whitelaw, the Home Secretary, to all intents and purposes the man who was in charge of the whole operation. For the SAS anti-terrorist team events had become a great deal more stable. By now access had been made into No 14 and efforts were being made to penetrate the dividing wall. To aid this inevitably noisy process, various sound distractions were supplied by men working on the underground gas supplies in the area. On the other side of the Iranian Embassy, the Ethiopian Embassy at No 17 was also fully cooperating.

By this time the basic plan had been formalised. This established that SAS men were to attack each floor at the same time, which helped to clarify areas of demarcation and the avoidance of overshoot. A mock-up of the floor layouts were constructed from timber and Hessian cloth, and then assembled in the Regent's Park barracks.

All morning threats were issued about executing the hostages

The police had adopted a softly, softly negotiating approach and thus managed to drag the siege out for several days. This additional time was desperately needed for the SAS to carry out covert reconnaissances, study plans, build models and, most importantly, locate the positions of the hostages and terrorists within the embassy building itself. A major break was the information gleaned from the released hostage Chris Cramer: in his debrief to the SAS, Cramer gave precise and detailed information about the situation inside the embassy.

By the sixth day of the siege – 5 May – the terrorists were becoming frustrated and the

Right: Pump-action shotguns are used in hostage-rescue operations, but only to blast out door bolts and hinges.

situation inside the embassy began to deteriorate. All morning threats were issued about executing hostages, and at 13:31 hours three shots were heard. At 18:50 hours more shots were heard, and the body of the embassy press officer was thrown onto the pavement. Immediately the police appeared to capitulate, stalling for time, while the SAS plans to storm the embassy were advanced. At this stage the police negotiator worked hard to convince the terrorist leader not to shoot any further hostages and that a bus would arrive shortly to take them to the airport, from where they could fly to the Middle East. As the telephone conversation took place, the SAS teams had taken up their start positions.

A hand-written note now passed control from the police to the SAS. Shortly after this, and while a negotiator from Alpha Control talked to Oan, the SAS moved in. Oan heard the first crashes and complained that the embassy was being attacked (this conversation was recorded, and one can clearly hear the stun grenades going off and then Oan's conversation being cut short by a long burst of machine-gun fire). For the SAS assault team, the waiting was over. The 'Go, go, go' command, guaranteed to get their adrenalin pumping, was given.

The assault came from three directions, with the main assault from the rear

At 19:23 hours, eight men abseiled down to the first-floor balcony using ropes secured to the embassy roof. The assault came from three directions, with the main assault from the rear. Frame charges were quickly fitted to the windows and blown, securing access to the interior of the building. Stun grenades were thrown in advance of the assault teams and the SAS went into action. Systematically, the building was cleared from the top down, room by room. The second-floor telex room, which housed the male hostages and three of the terrorists, was the priority target. Realising that an assault was in progress, the terrorists shot and killed one hostage and wounded two others before the lead SAS team broke into the room. Immediately, the black-clad troopers shot the two gunmen who were visible, but the third hid among the hostages. As rooms were cleared, hostages were literally thrown from one SAS soldier to another, down the stairs and out into the back garden of the embassy. At this stage they were all laid face down on the ground while a search was conducted for the missing terrorist (he was later found hiding among the hostages – all the other terrorists were killed).

Breaking the siege took just 17 minutes. The SAS took no casualties other than one man who got caught up in his abseil and was badly burned, and soon after affairs were handed back to the police. Meanwhile, the SAS vacated No 14 and went back to the barracks in time to watch themselves on TV. Still dressed in assault gear and clutching cans of Fosters lager (someone was on the ball!), they crowded around, eager to see themselves in action. Halfway through, Prime Minister Margaret Thatcher arrived to give her personal thanks to the boys. She circulated, as one man put it: 'like a triumphant Caesar returning to the Senate', her face glowing with pride and admiration at her Imperial Guard. Then, as the news started to show the full event, she sat down on the floor amid her warriors.

For the SAS it was back to Hereford and yet more training, but the British pride in the Regiment's achievement still endures. For the boys it was a time to let their hair down and relax a little. A party was organised, the highlight of which was the comedian Jim Davidson, dressed in full anti-terrorist gear doing a party piece.

RESCUING OUR COMRADES

SAS soldiers always look after their own. This means never abandoning individuals when in danger, and always supporting other Sabre Squadrons in times of need.

There are many individual accounts of SAS men in trouble and being rescued by their comrades, but by far the greatest of these episodes occurred in the small coastal town of Mirbat in Dhofar province during the Regiment's war in southern Oman (1970-76). In brief it is best summed up by the certificate associated with the David Shepherd painting that was commissioned some years after the event. Text from the painting of the Battle of Mirbat by David Shepherd includes the following words:

'At dawn on the 19th of July 1972, a large rebel force, about 250 strong, attacked the Port of Mirbat in the Dhofar Province of Southern Oman.

'When the battle started Corporal Labalaba and Trooper Savesaki, both members of a nine-man Special Air Service Civil Action Team, went to man a 25-pdr gun just outside the walls of the fort due North West of the town. Gunner Walid Khalfan of the Oman Artillery was already there. It immediately became clear that the main enemy

Left: An SAS 81mm mortar position on the Jebel Dhofar in late 1972. A similar position played a vital part in the epic Battle of Mirbat on 19 July 1972.

Left: The Dhofar Gendarmerie Fort at Mirbat, which was the main target of the ferocious *adoo* attack during the battle.

thrust was being directed against the fort, and in particular against the gun which was firing at point-blank range over open sights. Before long the entire crew were wounded. Captain Kealy, the Commander of the Special Air Service detachment, and Trooper Tobin, who was a trained Medical Orderly, then ran under fire from the main Special Air Service position to help save the gun.

'The rebels continued to attack with great ferocity and made repeated attempts to take the gun, often from within grenade-throwing range, and despite the supporting fire from the other five Special Air Service soldiers. The action lasted nearly four hours before a relief force and an accompanying air strike drove off the enemy. During this action Corporal Labalaba was killed, Trooper Tobin fatally injured and Trooper Savesaki and Gunner Walid Khalfan both seriously wounded.

'The official report records that the fate of Mirbat and of its occupants during the battle depended wholly on the resolve of the Civil Action Team. But for the action of these nine men, and particularly the leadership of Captain Kealy, the town would have undoubtedly fallen.'

To mass such a body of men was in itself a major achievement for the *adoo*

This was the last great attack by the *adoo*, the rebels of the People's Front for the Liberation of the Occupied Arabian Gulf (PFLOAG) in the Oman war, and they had planned it well (they had been losing ground; the assault was designed to show the people their continuing power). Only by a single stroke of bad luck had they miscalculated, for by coincidence there were two SAS squadrons in Oman at that time due to an end-of-tour handover. This simple fact, combined with the professionalism of the Sultan of Oman's Air Force (SOAF), truly upset the *adoo*'s plans. Had the result been different, the SAS war effort would have suffered a severe setback. The skill of the *adoo*'s planning and preparation were surpassed only by their security. To mass such a body of men, together with heavy support weapons, and bring them out onto the coastal plain undetected, was in itself a major achievement for the *adoo*. They arrived under cover of darkness, almost an hour before the first grey skies would filter through the monsoon mist. It was this seasonal cover of low cloud upon which they relied, not just to protect themselves from SOAF warplanes, but also to provide a clandestine approach.

The *adoo* had opened up with several artillery pieces

At around 05:00 hours the picket at the top of Jebel Ali, a small hill 1000m (1095 yards) to the north of Mirbat and halfway towards the Jebel Dhofar, was being manned by a section of Dhofar Gendarmerie (DG). The Jebel Ali is a dominating feature covering the town and the surrounding coastal area. The DG section was the first to be taken out. Surrounding the location to make sure there were no escapees, the *adoo* attacked, stealthily at first as they cut the throats of those still asleep and then, as the alarm was raised, opening up with small arms.

The exchange of fire was heard by those in the British Army Training Team (BATT) house. It was soon to be followed by 'thump, thump, thump' as the *adoo* mortars kicked in. Inside the BATT house, this chorus of explosions gave little cause for alarm as the enemy often lobbed in a few mortar rounds each morning. The first rounds had been off target, but by the third salvo they were shaking off the plaster inside the BATT house. Quickly the boys ran up the makeshift stairs leading to the roof. Here a defensive barricade of sandbags had been erected to form two gun-pits: one for the GPMG in the northwest corner, and one for the 0.5in Browning in the northeast corner. Just in front of the house was an 81mm (3.2in) mortar pit. Captain Mike Kealy, the BATT commander, and his men observed the fire coming from the direction of the Jebel Ali. They also saw several streams of tracer fire lace their way over the house and into the town. Almost at once, several more mortar bombs fell close to the

Dhofar Gendarmerie (DG) Fort situated on the northern side of town. In addition, the *adoo* had opened up with several artillery pieces, and a crushing barrage descended on the town. As the shells exploded, the *adoo* rose in waves and advanced towards the DG Fort.

The BATT house took more incoming rounds but the SAS men, although ready to defend, held their fire. Among them were some of the most experienced soldiers the Regiment had, men who had already seen their share of action. To them this sounded nothing more than a stand-off attack, and so the first thing was to assess the situation. From the roof of the BATT house, Captain Kealy could see the firefight raging on top of the Jebel Ali. Turning,

he then checked to see what damage had been done to the town itself. He then shouted orders for the 81mm (3.2in) mortar to open fire in support of the Jebel Ali, while the rest of the SAS men took up their positions behind the sandbagged emplacements and waited for targets. The amount of incoming fire worried Kealy, and as a safeguard he ordered his signaller to establish communications with SAS headquarters at Um al Gwarif. Additionally, the big Fijian, Trooper Labalaba, left the house and ran the 500m (1635ft) to the DG Fort, where he manned an old World War II 25-pdr artillery piece. By 05:30 hours, it was light enough for Kealy to make out the silhouette of the gun position and the DG Fort just to its east.

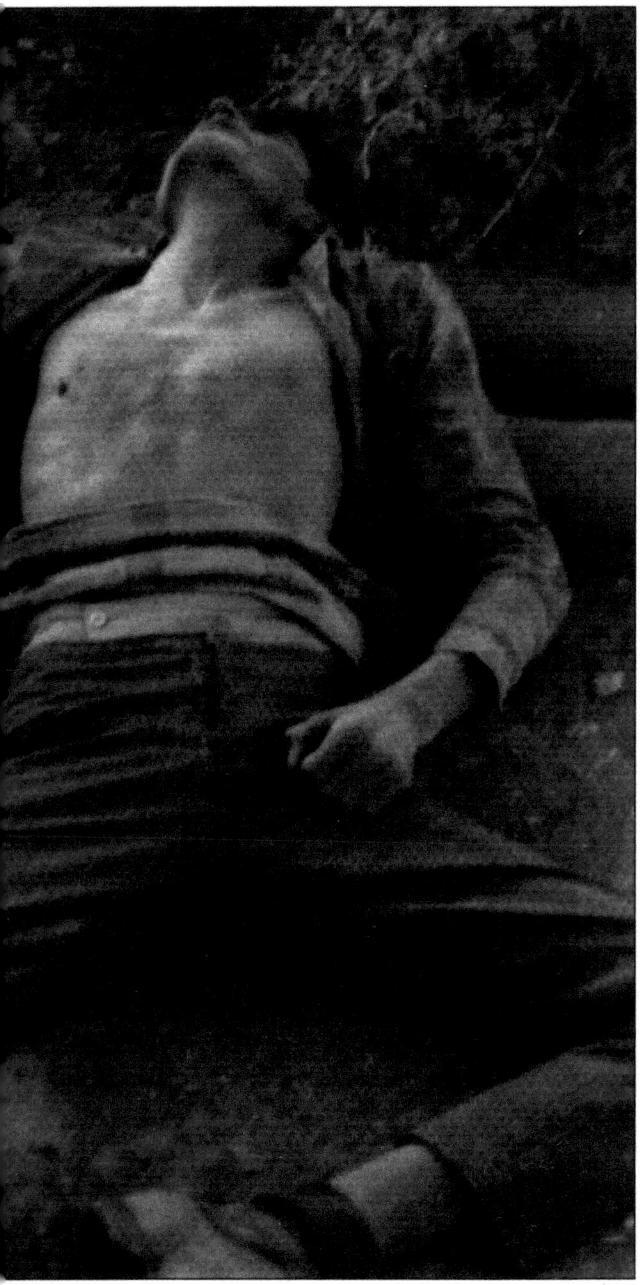

Despite the ferocity of the attack, at this stage there seemed no reason to think that this was anything more than a strong stand-off attack. However, the duty signaller at Um al Gwarif was informed and told to listen out. For the rest of the guys back at base, they were happily getting showered and making ready for the ranges. Most of these were from G Squadron. Their signals sergeant, 'Ginge' Reese, passed comment over breakfast about the attack: 'Mirbat are still taking a hammering, its a bit long for a stand-off.' Five minutes later, 'Duke' Pirie, the B Squadron commander, started to organise a relief force.

Suddenly, a vast amount of small-arms fire started pouring into the town

At Mirbat most of the *firqat* was out on patrol, leaving the town lightly defended, the houses manned by old men, women and children. Only a few *firqat* members had remained behind, and these, together with a group of Dhofar Gendarmerie, manning the old fort (the Wali's Fort), were the only help the SAS could call on.

Around Mirbat, *adoo* shell fire continued to increase. Suddenly, a vast amount of small-arms fire started pouring into the town. Through the mist, figures could be seen approaching the perimeter wire from the direction of the Jebel Ali. At first there was some hesitation as to their identity; it was possible that the *firqat* had returned. Then the advancing figures opened fire – they were *adoo*.

As battle was joined both SAS machine-gun bunkers opened up, and at the same time the

Left: Dead adoo soldiers after the Battle of Mirbat. The defeat was regarded as a loss of face for the enemy, which in an Arab society is regarded as a grave setback.

81mm (3.2in) mortar increased its rate of fire. In the gun-pit by the DG Fort, the 25-pdr gun sent shell after shell into the massing *adoo*. The battle flowed back and forth, then a radio message came through on the Tokki (a small commercial walkie-talkie used by the SAS throughout the Oman war) from Labalaba, saying that he had been hit in the chin while operating the 25-pdr gun. A man of such stature is not given to reporting such trivia, and those at the BATT house suspected he was badly injured. Immediately Captain Kealy sent Labalaba's Fijian countryman, Savesaki, to his aid. The gunners provided supporting fire from the roof of the BATT house as they watched Savesaki run the gauntlet of tracer and exploding shells. Zigzagging, he ran as if on the rugby field and going for a try. He finally dived headlong into the gun-pit.

The *adoo* looked calm and seemed disciplined moving forward

Savesaki found Labalaba firing the gun on his own. The big Fijian gave no indication that he was injured. Instead he pointed to the unopened ammunition boxes, and the desperate need to keep the gun firing. Much of the *adoo* attack was now directed against the gun itself, and its crew needed more help. For a brief moment Savesaki left the pit to solicit help from the DG Fort a few metres away. As the rounds zipped past his head, he banged on the fort door until at last he was heard. The first man to appear was the Omani gunner, and Savesaki grabbed him. Together both men raced the few metres back to the gun-pit. Savesaki cleared the sandbags, but the Omani gunner fell forward as a bullet ripped out his guts.

It was now light enough to see groups of men near the outer perimeter fence that protected the three open sides of the town. Behind them, wave after wave could be seen advancing towards Mirbat in support. At that moment several rockets slammed into the DG Fort, causing great chunks of masonry to be blown from its ancient walls. The *adoo* looked calm and seemed disciplined, moving forward in sections of about 10 men. From all around

Right: An SOAF Strikemaster jet. The timely intervention of the Strikemasters saved the day for the SAS under attack at Mirbat.

whistles could be heard driving the *adoo* on. Suddenly another mighty blast rang out, and the front of the fort exploded in a cloud of dust, totally obliterating it from sight. As vision returned, the SAS on the BATT house roof focused on a new threat. The *adoo* were at the perimeter wire. Swarms of young men threw themselves headlong onto the barbed, razor steel – as one died another took his place. Men could be seen all along the wire, but the main breakthrough seemed to be in front of the fort. At this stage most of the enemy fire was concentrated on the fort in general and at the gun-pit in particular. Several more heavy explosions were heard, and once more the old fort disappeared in a cloud of dust (it was discovered later that apart from the recoilless guns and mortars, the *adoo* had a Carl Gustaf rocket launcher that had been stolen from British forces in Aden some years earlier).

The *adoo* were now inside the perimeter and advancing on the fort and gun position in large numbers. In support of Labalaba and Savesaki, the men at BATT house brought all their guns to bear. When the *adoo* still outside the wire started to realise where all the return fire was coming from, they swiftly retaliated.

His shirt soaked in blood, Savesaki propped himself against the sandbags

(Author's Note: There is a corroborated story that as the *adoo* penetrated the wire perimeter in the early stages of the assault, a lone figure could be seen driving them on. He was described as being very tall and extremely well dressed in full Chinese army uniform, with courage to match. In full view of the horrendous firefight that surrounded him, he stood proudly urging his men on – until a bullet took him down.)

The gun, now levelled directly at the wire, boomed as Labalaba fired point-blank into the charging figures. Both Fijians moved in unison, Savesaki passing the rounds and Labalaba

slamming them home, before blasting them into the wall of humanity just a few metres away. Suddenly Savesaki cried out 'I'm hit', and slumped back against the sandbags. Without a loader the gun fell silent. His shirt soaked with blood, Savesaki propped himself against the sandbags, grabbed his SLR and continued to fire; Labalaba made a quick grab for a small 60mm (2.36in) mortar that lay close by. He almost made it, but then a bullet took him in the neck. The mighty mountain of a man, a true gladiator, fell dead.

Communications with the gun-pit had been lost, and Captain Kealy decided that he and an SAS medic, Trooper Tobin, would risk going forward to give assistance. Before they left, Kealy contacted Um al Gwarif, informing them that things were not going too well and that air cover was desperately needed. He also requested a helicopter to take out Labalaba. Additionally, if the firefight continued at its frenetic pace more ammunition would be required. It is a little-known fact that if an SAS man gets into serious trouble he requests the 'Beast'. This call alerts the headquarters unit to the seriousness of the situation, and every SAS soldier who can carry a weapon is pressed into service. In the case of Mirbat, the Beast had been put on standby. Most of those mustered were from the newly arrived G Squadron. The men were already dressed and equipped for the range, all that was required was a little extra firepower. It was customary for senior troop leaders to go off to the locations which they would be taking over. While they were doing this, the rest of the squadron would unpack stores and test-fire their weapons on the range. Such was the case on the morning in question. As 'Ginge' Reese

ROYAL IRISH RANGERS
WHO DARES WINS
23892771 SERGEANT
T. LABALABA. B.E.M.
THE ROYAL IRISH RANGERS
22 SPECIAL AIR SERVICE REGIMENT
19TH JULY 1972 AGE 30 YEARS
GREATER LOVE HATH
TO A
DEAR

Left: The tombstone of the Fijian, Labalaba, engraved with the insignia of his regiment in Saint Martin's graveyard, Hereford.

spread the news throughout the cookhouse, most of the men stopped eating and ran for the armoury. It took about five minutes for 22 of them, under the command of Captain Alastair Morrison, to get together an impressive array of weapons. Eight GPMGs and several grenade launchers were among the group – the enemy was going to know about it when these men entered the fray. The total ammunition count for the reinforcements was over 25,000 rounds.

Back at Mirbat, Captain Kealy and Trooper Tobin worked their way forward to the gun-pit. As they approached the fighting increased and both dived for cover. Luckily for them, there was a shallow wadi running in their general direction, and this afforded them a measure of protection. With a final dive Tobin rolled into the gun-pit. Kealy was about to follow, but realising there was not enough room, and tripping over the body of a dead gendarme, he threw himself headlong into the sandbagged ammunition bay.

Nearby the Omani gunner moaned as he clutched the wound in his stomach

Tobin could not believe the mess. Labalaba lay face down and was very still; Savesaki sat propped against the sandbags, SLR still in hand; and nearby the Omani gunner moaned as he clutched the wound in his stomach. Assessing the priorities, Tobin quickly set up a drip on the severley wounded Omani gunner. Savesaki was badly wounded in the back, but despite the severe loss of blood he continued to fight, covering the left side of the fort. The firefight had reached its height, and the *adoo* made a real effort to overrun the gun. As Kealy concentrated amid the mayhem, he saw *adoo* close-by the fort wall. They threw several grenades, which bounced by the lip of the gun-pit before exploding. An *adoo* appeared at the side of the gun-pit but Kealy cut him down. In the pit, Tobin reached over the inert body of Labalaba and then, realising there was little he could do, made to move away. At that moment a bullet took away half his face. His whole body stiffened, then he fell by the side of the big Fijian, mortally wounded. The gun-pit seemed done for. Kealy, rapidly running out of full magazines, reloaded quickly, keeping his head below the sandbags. Suddenly there came an almighty explosion – the SOAF jets had arrived. Kealy called into the Tokki, managing to make contact with the BATT house: 'Laba's dead. Tak and Tommy both VSI. Get help.' Back at the BATT house they heard the sound of an aircraft, and thinking it a casevac helicopter one of the soldiers went to see it safely landed. Then

Below: The tombstone of Trooper Tobin, who was killed at Mirbat going to the aid of his Fijian comrade, Labalaba.

he realised it was not a chopper but a BAC Strikemaster light attack aircraft. The soldier ran back to the roof, and snatching up his SARBE (Surface-to-Air Rescue BEacon) radio, directed the pilot with precise information regarding enemy forces.

Despite the monsoon weather, the pilots had dived out of the cloud just feet above the ground. There before them was the enemy. Their machine guns roared as the first two jets made pass after pass, driving the *adoo* back into a large wadi outside the perimeter wire. As Kealy silently thanked the pilots, he saw the large 227kg (500lb) bomb unhook itself from one of the Strikemasters and fall into the wadi where the *adoo* had taken refuge.

(Author's Note: the SOAF pilots did a fantastic job, streaking over the DG Fort and keeping the *adoo* from any further penetration through the wire perimeter. Time after time they dived from the low cloud to run the gauntlet of ground fire that was pouring

Left: Omani dwellings near the Shirshitti caves, scene of an SAS and Omani operation in January 1975.

towards them from the ground. I was at the airport ready to fly out with an 81mm (3.2in) mortar, and the whole battle was broadcast live over the intercom. We had a similar system at Um al Gwarif.)

'I'm hit, I'm hit,' called the pilot's voice. 'No fuel. Repeat, no fuel.' Seconds later, as the fire trucks raced across the tarmac, the jet could be seen lining up for the runway, its tailpipe bleeding black smoke. The pilot literally slammed the jet into the ground as fire tenders, which had raced side-by-side with the aircraft, pumped foam to extinguish the fire.

By this time the desperate situation was clearly understood at Um al Gwarif. The G Squadron relief force had already loaded themselves into three choppers, and was rapidly heading down the coast. The chopper covered the 50km (31 miles) to Mirbat in about 10 minutes. Because of the cloud, the men of the relief force were dropped off to the south of Mirbat and instantly made contact with an *adoo* patrol that was covering the rear. The *adoo*, consisting of one older soldier and three youths, were held up in a cave and refused to surrender. Several 66mm (2.6in) LAW rockets slammed into the entrance, followed by fire from several GPMGs, and in this fashion the *adoo* picket was quickly neutralised.

Slowly the two men covered the bodies and gave what comfort they could

With the jets taking the sting out of the *adoo*, Kealy had time to crawl forward and examine the gun-pit. He could see that the Omani gunner was still alive, and so too was Tobin although his wound looked horrendous. Savesaki lay listless against the sandbags, his whole body seemed covered with blood, but he still managed a smile. Slowly the two men covered the bodies and gave what comfort they could. The SOAF jets pushed the *adoo* back, and were now dropping 227kg (500lb) bombs on the Jebel Ali. Kealy and Savesaki tried to shout at the Gendarmes in the fort to open the door, but it was to no avail. Then the SAS relief force arrived. Moving through the town had been difficult (as with the SAS in Mirbat, identifying who was *adoo* in the monsoon conditions was not easy). The G Squadron men had to be alert, as *firqat* members had positioned themselves in groups of twos and threes around the town.

As the mess started to clear and the *adoo* withdrew, harassed all the way by the

Strikemasters, concern was raised about the main *firqat* patrol that had been lured away before the attack. Obviously, having heard the amount of gunfire coming from the direction of Mirbat, they would immediately return to their base. The problem now was whether or not they would they run headlong into the retreating *adoo*.

Although several of the choppers had been hit, they continued to ferry in more reinforcements, extracting the wounded in return. Tobin and the Omani gunner, both seriously wounded, were 'casevaced' on the first available flight. Savesaki, who suffered wounds from which a normal man would have died, walked calmly to the chopper without assistance. Three young *adoo*, captured and held in the BATT house, were also sent back for interrogation. Meanwhile the relieving force commander, Alastair Morrison, reorganised Mirbat's defences and, with the aid of two Land Rovers, started to collect the dead and wounded

adoo. The final count of dead *adoo* was 38. They were stacked neatly in the back of a Skyvan transport and flown to the southern capital, Salalah, where they were laid out for the whole population of the city to see.

The Dhofar war was never the same after the Battle of Mirbat. The *adoo* had given it their best shot and failed, but only just. The chances that a second SAS squadron would be just 50km (31 miles) away at the time of the attack was something they could not have foreseen. Additionally, they underestimated the expertise and nerve of the SOAF pilots. Two years later the SAS fought another great battle in Oman: the Shirshitti Caves operation.

This took place in 1974, more or less as the war was coming to an end. Major General Creasey, commander of the Sultan of Oman's Armed Forces, ordered the Iranian Battle Group (an Iranian special forces battalion had been sent to the support of the Omani forces by the Shah) to advance from the air base at Manston and secure the coastal town of Rakyut, some 27km (17 miles) to the south. Its mission was to clear the *adoo* stores complex located in the caves of the Shirshitti Wadi. These caves were said to hold tons of weapons, food and combat supplies. Their capture would help bring the war to a successful end.

As the lead elements approached Zakhir Tree they met serious resistance

The attack started in mid-December but did not get very far. The *adoo* had seen the Iranians coming, and in heavy fighting inflicted severe casualties on them. Unable to sustain the losses, the Iranians had called off the attack. At this stage of the war, though, defeat was not an option for the Omani government. Almost immediately a decision was taken by Creasey to launch another attack using Omani forces, SAS and *firqat*. There was a problem, however, in the fact that all the Omani regiments in the south were already hard pressed with prior commitments. So it was that the Jebel Regiment was flown down from northern Oman. After some swift training it was sent into battle.

By 4 January 1975, the force was ready. The plan was fairly simple: seize an old airstrip called Defa in order to establish a supply point, and then secure the ridge that overlooked the Shirshitti depression in which the *adoo* cave dumps lay. As always, the SAS and *firqat* led the advance. Defa was quickly taken and the advance rapidly continued. As the lead elements

Left: SAS troops during the Shirshitti caves operation. The author is in the front row on the right of the photograph.

Left: The Shirshitti Wadi, which was the scene of especially fierce fighting between the *adoo* and the SAS.

approached a landmark known as the Zakhir Tree, they met serious resistance. For some reason the *firqat* did not perform well, but the SAS men laid down a thick and furious fire. By mid-afternoon they had managed to reach a clearing called Point 985, whereupon a base was established. During the night the *adoo* attacked at very close range, killing four Omani soldiers and severely wounding many more. At times it was difficult to establish the locations from where the *adoo* fire was coming, and as a result the Omani soldiers in the perimeter defences became extremely agitated, and fired more for relief than to hit a specific target.

Next day the force advanced down into the Shirshitti. By mid-morning the regiment's Red Company had reached the Shirshitti Wadi, but the commander realised that he had moved too far south, navigation in the thick bush being difficult. There also seemed some confusion as to the location of the other two companies. At this stage most of the SAS men had attached themselves to the various command headquarters. With Red Company was Lance-Corporal Thomas. As the lead platoons broke cover (against the advice of Corporal Thomas) into an area clear of bush, the *adoo* opened fire. Within seconds most of the platoon's men were dead, cut down by the ferocious *adoo* firepower.

It was only at gunpoint that order could be restored among the units

The company commander and several other men rushed forward to get a better look at the tactical situation. The *adoo* were waiting, though, and cut down these men, too. Even to the hardened SAS men, it became clear that the situation was getting out of control. As in such circumstances, they quickly grouped together for support. All around them, though, the surviving Omani soldiers dropped their weapons and ran, despite the desperate efforts of their British officers, many of whom were killed, to restore control.

It is not written in official sources, but it was only at gunpoint that order could be restored. In order to regain control of the situation, the SAS party with Red Company headquarters called in an air strike. However, such was the confusion on the ground that the first strike mistakenly hit the company headquarters, one of the rounds striking an SAS soldier in the back. Fluorescent air-marker panels were hastily pulled from belt kits and wrapped around the shoulders for recognition. Eventually, with massive firepower from artillery, mortars and Strikemasters, the *adoo* were driven back. Amid all this carnage there were several individual acts of great courage, as men braved the horrendous fire to rescue friends and comrades.

It was decided to blast the *adoo* out of their stronghold

Although the dead were left behind, all the wounded and weapons were recovered before a tactical withdrawal was ordered. As the shaken troops made their way back to Point 985, shots could be heard coming from the wadi – the *adoo* were confirming their kills. To alleviate this gruesome sound, a full-blown mortar barrage was called down on the battle area.

That day the *adoo* had won a victory, but they were later to pay the price. As the defences at Point 985 were reinforced, it was decided to blast the *adoo* out of their stronghold. So it was that for each minute of the following two days every weapon that could reach the Shirshitti Wadi was fired. Artillery, armoured cars, mortars and warplanes were involved, and a bombardment was even delivered by an Omani warship lying off the coast. The air thundered with high explosives, and the dust and phosphorus smoke hung over the Shirshitti Wadi like a permanent cloud.

(Author's Note: At this stage both SAS mortar men controlling the baseline at Point 985 had been wounded, and I was withdrawn to replace them. There were six mortars in two groups of three. Each half hour it was our turn to fire, each tube firing 10 rounds in a mixture of white phosphorus and high explosive. In the

Right: The Argentinian submarine *Santa Fe,* which was damaged during the British retaking of South Georgia in April 1982.

next few hours the mortars alone consumed 8000 rounds. Even with this massive barrage, the *adoo* still found time to carry out night attacks on our position, and several large rockets were also fired into our camp, with devastating effect. Amid the carnage taking place in the wadi, the *adoo* also launched three giant rockets. I watched in disbelief as the first one flew over our position, a great flame burning from its tail. When it fell to earth the whole world seemed to shake. Luckily for us only one hit the camp, though rather unluckily it hit the ammunition bunker.)

In the end the Shirshitti caves were taken, and vast *adoo* stores were captured. And although the Battle of Mirbat is the most famous engagement of the Oman War, such actions as the Shirshitti caves will remain in the memory of those who were there.

The cold and wind were so bad that they would not have lasted 24 hours

Rescuing comrades also occurred during a later conflict: the 1982 Falklands War. One such incident concerned Captain John Hamilton, who was troop commander of D Squadron's Mountain Troop, and as such had been in the Falklands from the very start. One of the troop's first tasks had been to set up observation posts (OPs) on the island of South Georgia, with the view to its recapture from the Argentine garrison as the first step towards retaking the Falklands themselves. The first insertion of SAS fell to men of Hamilton's troop, when on 22 April they were dropped by helicopter onto Fortuna Glacier. As every SAS man will tell you, a feature on a map is not the same as that feature on the ground. Almost immediately the Mountain Troop members realised that to stay where they had been dropped would mean certain death. The cold and wind were so bad that they would not have lasted 24 hours, added to which it was impossible to see more than a few metres. Thus the next morning an immediate recall was requested. The extraction cost two Westland Wessex helicopters, both of which crashed as a result of the weather. In what can only be described as superb flying, a third chopper, overladen and in blind conditions, managed to extract every man.

The SAS did not abandon the idea of establishing OPs, but decided to use a different method of insertion: by boat. This also met

with several problems, as in difficult weather and poor visibility the two boats initially missed South Georgia altogether. One finally made land, and the SAS men lay huddled on the freezing shore for two days. The other boat, which as a result of engine trouble had drifted out to sea, was rescued by the Royal Navy. Those dinghies which did reach the island succeeded and established several OPs. The main Argentine force on the island was stationed at Grytviken, and comprised about 100 men made up of Marines and the crew of the submarine *Santa Fe*, which was in the harbour delivering supplies. On 25 April the submarine was spotted leaving. Several British helicopters attacked and damaged it, forcing the submarine to return to Grytviken harbour. At the same time a British force made up of Special Boat Service (SBS), Royal Marines and SAS assaulted the base. They were inserted by

Right: A wrecked Argentinian Pucará on Pebble Island following the SAS raid on the night of 14/15 May 1982.

helicopter and supported by covering fire supplied by the Royal Navy. By late evening the Argentine garrison had surrendered, as had a small detachment at the old whaling station at Leith. It was the first victory of the war and a massive propaganda boost for the UK.

After this, Captain Hamilton took part in the raid on Pebble Island, off West Falkland, when, on the night of 14 May, 45 men of D Squadron attacked the Argentine airstrip located there. During this action they destroyed six FMA IA 58 Pucara attack warplanes, four Beech T-34C Turbo-Mentor armed training aircraft and one Shorts Skyvan transport. Additionally, they killed several pilots and ground crew and denied the enemy further use of the airstrip. The raid was infiltrated and extracted by helicopter, and was a classic example of an SAS hit-and-run mission.

The four SAS men of the patrol desperately tried to fight their way out

Towards the end of the war, on 10 June, a patrol led by Captain Hamilton was spotted close to Port Howard on West Falkland. The four SAS men of the patrol desperately tried to fight their way out. As two withdrew, Hamilton and his signaller gave covering fire. During the firefight Hamilton was hit, but he continued to provide supporting fire. Some time later the overwhelming force of Argentines killed him and captured his signaller – a sad loss. However, Hamilton and his comrade, through their unstinting efforts, had won sufficient time for the rest of the patrol to escape.

In more recent times of trouble, such as the Gulf War (1991), there emerged new stories of rescue. *Bravo Two Zero* and *The One That Got Away* are somewhat differing accounts of a patrol isolated in the middle of the Iraqi desert and being hotly pursued by the enemy. The patrol had been inserted by helicopter deep behind enemy lines. Its mission was to observe the movement of Iraqi 'Scud' missile-launch vehicles, whose activities were threatening to bring Israel into the conflict. Almost immediately the SAS patrol was discovered by the enemy and engaged in a firefight. The unit then made a run for the Syrian border. Hunted every inch of the way, travelling in highly adverse weather and terrain conditions, they killed over

200 Iraqi solders as they fled. The outcome was that one man made it, three died and four were captured. The last were tortured, but survived to be released at the end of the war.

These are just a few of the rescue events in which SAS men have risked, or given, their lives to help their comrades. There have been many other individual accounts since the Regiment's creation in North Africa in World War II, and doubtless there will be many more in the future. On operations SAS men trust each other implicitly; this trust is the cornerstone that sets them apart from other elite soldiers and upon which their success is based.

RESCUING SOCIETY

When society is in danger, both at home and abroad, the SAS stands ready. In the 1991 Gulf War, for example, the Regiment helped save Kuwait from the tyranny of Saddam Hussein.

There are many accounts of the SAS being despatched to rescue society. These episodes, encompassing the world and having taken many different forms, have ranged from helping the tiny country of Belize, threatened by neighbouring Guatemala, to the prevention of a car bomb attack designed to kill soldiers and civilians in Gibraltar. Some SAS actions reach the daily newspapers, but others remain secret.

The first recorded European settlement in Belize was begun by shipwrecked English seamen in 1638. Over the next 150 years more English settlements were established. This period was also marked by piracy, indiscriminate logging and sporadic attacks by Indians and the neighbouring Spanish settlements. The area became the crown colony of British Honduras in 1862. Several constitutional changes were much later made to expand representative government. Full internal self-government under a ministerial system was granted in January 1964, the official name of the territory was changed to Belize in June 1973, and full independence was granted on 21 September 1981. Yet the British Army remained responsible for the protection of Belize, and accordingly a British garrison

Left: Two heavily laden SAS Land Rovers and their crews behind Iraqi lines in February 1991 during the UN Gulf War against Iraq.

remained *in situ*. This force was designed primarily to deter military advances from neighbouring Guatemala, which is much larger and lays claim to Belize. From time to time the Guatemalans have rattled their sabres and begun preparations to take Belize by force. Much of their rhetoric was taken with a pinch of salt, but on one occasion they seriously intended to invade. The 1800-strong garrison was rapidly reinforced in November 1975. The author takes up the story:

'It was one of those quick-move jobs. One minute we were happily doing a spot of training in Hereford, and the next we are winging our way over the Atlantic. The faithful Lockheed C-130 Hercules transport took off from Brize Norton bound for Belize, with a refuelling stop at Nassau in the Bahamas. Here we were given permission to leave the aircraft and stretch our legs while the transport was being serviced. It was at this stage that I noticed the RAF loadmaster beating one of the engines with a

Left: Rescuing society from drug barons. SAS soldiers in Colombia as part of the war against international drug traffickers.

broom. About 10 minutes later we were informed that the flight had gone US (unserviceable) and that we would have to spend the night in Nassau – I could have kissed him! If you have ever wondered why the sailors of old never returned from these islands, one breath of the relaxing air will explain everything. Forgetting the war, we were taken by a fleet of taxis to a Trust House Forte hotel. Here the whole squadron was given grand rooms, many right on the beach. "Eat what you like, gentlemen. For you happy-hour has just begun," said the manager as a way of saying that it was all free.

'Unfortunately, next morning the Hercules was fixed and we continued our journey to Belize, most of us nursing a hangover. We arrived about 14:00 hours in the afternoon, and quickly set about equipping ourselves for the jungle. I was given a seven-man patrol. This was a bit large for the jungle, but it suited our task. We were to penetrate the thick jungle in central Belize and move up to the border with Guatemala. By 16:00 hours we were ready to get choppered in, and that is when this guy arrived from the government. He had come to tell us about the type of jungle in Belize. To a man the squadron sat on the floor as Major Rose introduced him. The first words he said were "fer de lance". We looked at each other in some amusement, wondering what the hell was he going on about. He then proceeded to tell us all about this snake that could fly. It would leap out of the trees and bite your neck, and then you died. Charming, and on top of this we still had to take care of the Guatemalans.

'The boys behind me had gone directly into combat mode'

'We infiltrated at dusk, and after clearing the drop-off found a spot to hole up for the night. To be fair the boys went into jungle mode straight away: no talking, no cutting, everyone alert. Next day we set off, and as two of the guys were new to the troop I decided to carry out one or two contact drills. This we did, and I am confident that had we hit the enemy the patrol would have held its ground and extracted in good order. Three days later, as we neared our objective, I suddenly saw the lead scout, Tony, move to one side and bring up his weapon. I heard two loud bangs, followed by the cry "fer de lance". Initially thinking that this was a enemy contact, the boys behind me had gone directly into combat mode, ready to take on the worst. Two seconds later, as the words "fer de lance" pierced the air, you could not see them for dust. Fighting Guatemalans is one thing, flying snakes are another.

'Our task was to observe a main jungle route that led from the Guatemalan side and stretched over the border into Belize. It was thought the Guatemalans might use this to infiltrate troops along. In the event of invasion we were to report all enemy movements and strengths. If we were compromised, the plan was to bug out and run north to Mexico (agreement had been reached between the British and Mexican governments for the internment of all UK soldiers until the end of the war; blood money was issued to make this journey possible).

In the war against the Irish Republican Army the SAS has been relentless

'By day five, Hawker Siddeley Harrier STOVL (Short Take-Off and Vertical Landing) warplanes of the Royal Air Force had arrived. They flew down the border in teams of four, stopping now and then to hover, like giant angry bees waiting to sting. In full sight of any Guatemalan forces, they flaunted their massive firepower. It was enough, and a day later the Guatemalans called off any thoughts of invasion. Once more British gunboat diplomacy had worked, and Belize had been saved from the threat of war.

'We remained in the jungle for several more days carrying out "hearts and minds" activities before being extracted to Belize City and thence home. Before I left I purchased a T-shirt with "I fought the Guats" printed on the front and "ALL DAY" on the back.'

In the war against the Irish Republican Army (IRA), the SAS has been relentless in its hard-hitting attacks. Most of these have ended in the same way: a quick shoot-out resulting in the deaths of several terrorists. At times the SAS has been called upon to work on the British mainland against the IRA and also, on a very few occasions, abroad.

It was in March 1988 when information filtered through the security screen that the IRA was planning to explode a bomb in Gibraltar. The IRA team consisted of three people – two men, Sean Savage, Daniel McCann, and a woman, Mairead Farrell – each of them with a history of terrorist activity. The three, later acknowledged by the IRA as an active service unit, had been spotted in southern Spain by the Spanish and their presence reported to the British security services. They had been trailed for months, and many conversations between the three had been recorded. These provided information about the cell's target, which was the British garrison in Gibraltar, and the planned method of attack: a car bomb. As events became focused, the target was 'pinged' to be a military ceremony at which several military bands would be parading.

It was also known that the IRA had developed a new type of remote-control device which could detonate a car bomb from a distance. This would have a bearing on later events during the operation.

In late 1987 Savage, a well known IRA bomb maker, had been located in Spain, and with him was McCann. MI5, the British counter-intelligence organisation, spent the next six months watching the two, gathering vital information that seemed to confirm they were going to carry out a bombing. On 4 March 1988, Farrell arrived at Malaga airport and was met by the two men, and from that time onwards it seemed likely that the bombing was on. At this stage the SAS was invited to send in a team, the Regiment and the intelligence services having enjoyed a good working relationship since the mid-1970s.

The best spot to cause the most damage was thought to be the plaza

The Gibraltar police department was informed of the operation, and instructed that the IRA active service unit was to be apprehended. For a while, contact with the IRA cell was lost, but by this time the target had been defined. It was suspected that one car would be delivered onto the 'Rock' and parked in a position along the route taken by the military parade. This car would be 'clean', a dummy to guarantee a

Above: SAS soldiers in Belize in November 1975, as part of the deterrent against a Guatemalan incursion.

parking space for the real car bomb. The best spot to cause the most damage was thought to be the plaza in which the troops and public would assemble. This proved to be correct. At 14:00 hours on the afternoon of 5 March, a report was received that Savage had been spotted in a parked white Renault 5 car. There was a suspicion that he was setting up the bomb's triggering device. Not long after this, another report was received to the effect that Farrell and McCann had crossed the border and were making their way into town.

The SAS men were immediately deployed, and once Savage was out of the way an explosives expert did a walk past of the Renault. No visual tell-tale signs, such as the rear suspension being depressed by the weight of a bomb, were detected. However, if the IRA cell was using Semtex plastic explosive, 15kg (33lb) or more could be easily concealed without altering the balance of the car. After consultation it was considered that the car probably did contain a bomb. At this stage Joseph Canepa, the local police chief, signed an order passing control to the SAS. Operation 'Favius', as it was known, was about to be concluded. The orders given to the SAS men were to capture the three bombers if possible. But, as in all such situations, if there is a direct threat to life, either SAS or civilian, they hold the right to shoot. It was stressed that the bomb would more than likely be fired on a push-button detonator.

The SAS men, dressed in casual clothes, were kept in contact through small radios hidden about their persons, and each soldier was armed with a 9mm Browning High Power pistol. Savage linked up with McCann and Farrell, and after a short discussion all three made their way back towards the Spanish border. Four of the SAS team shadowed the trio. Suddenly, and for some reason that remains unexplained, Savage turned around and started to make his way back into the town. The SAS team therefore divided: two of the men turned to follow Savage, and the other two stayed with McCann and Farrell.

A few moments later fate took a hand. A local policeman, driving in heavy traffic, was recalled to the station. It was said later that his car was required, but whatever the reason he turned on his siren to expedite his orders. This action happened close to McCann and Farrell, making the pair turn nervously. McCann made eye contact with one of the SAS soldiers, who was no more than 10m (33ft) away. In response to this the soldier, who was about to issue a challenge, said in evidence later that McCann's arm moved distinctly across his body. Fearing that he might detonate the bomb, the soldier pulled his pistol and fired. McCann was hit in the back and went down. Farrell made a movement for her bag, and was shot with a single round. By this time the second soldier had drawn his pistol and opened fire, hitting both terrorists. On hearing the shots, Savage turned to be confronted by the other two SAS men. A warning was shouted this time, but Savage continued to reach into his pocket. Both SAS men fired, and Savage was killed.

The SAS soldiers were taken to court by relatives of the three IRA terrorists

As the first news of the event hit the media it all seemed to have been a professional job, but the euphoria was short-lived. No bomb was found in the car, and all three terrorists were found to be unarmed. Although a bomb was later discovered in Malaga, the press and the IRA had a field day. Allegations were made and witnesses were found who claimed to have seen the whole thing. According to some of the 'eye witnesses', the IRA trio had surrendered, their arms had been in the air, and then they had been shot at point-blank range after being forced to lay on the ground.

The SAS men were held up as killers. No matter that they had probably saved the lives of many people and dispatched three well known IRA terrorists, they would stand trial. In September 1988, after a two-week inquest, and by a majority of nine to two, a verdict was passed of lawful killing. Although this satisfied most people, the story did not end there. The SAS soldiers who had taken part in the shooting in Gibraltar were taken to court by relatives of the three IRA members killed. The European Commission of Human Rights in Strasbourg decided, by a majority of 11 to six, that the SAS did not use unnecessary force. They said that the soldiers were justified in opening fire as the IRA members were about to detonate a bomb. However, they did refer the case to the European Court of Human Rights. As a result of this court case, the British government was forced to pay compensation.

Michael Rose had that dash that made him stand out from his peers

The SAS role in Bosnia started in the early 1990s, when Lieutenant-General Sir Michael Rose, the United Nations (UN) commander, took on the formidable task of sorting out the Serbs and their aggression towards the Moslem population of Bosnia-Herzegovina (one of the states which emerged from the dissolution of the federal state of Yugoslavia). It was a job that had taken its toll on his predecessors: the Canadian General Lewis MacKenzie had been labelled a Serb-lover; the French General Philippe Morillon was totally ineffective; and the Belgian Lieutenant-General Francis Briquemont asked to be relieved six months early. Rose came from a different breed.

(Author's Note: I first knew Mike Rose as a G Squadron troop officer, and I would say this of him. In the early days he was fun; he had that dash that made him stand out from his peers. By the time he had made squadron commander we were in Armagh, Northern Ireland. He could really communicate with the boys: it was nothing for him to suddenly decide to go out on patrol acting as a trooper, leaving the corporal or sergeant in charge. He would listen, laugh, make suggestions, but he would never interfere.

As the Guatemalans started to rattle their sabres against the tiny British-protected state of

Above: The tombstone illustrates the IRA's view of Operation 'Favius', the SAS action to stop the terrorist bombing in Gibraltar.

Belize (see above), G Squadron, under Rose, was sent to intervene. I remember it well, for two hours before we went into the jungle he promoted me to sergeant. He simple pulled me to one side and said 'I'm making you sergeant and beefing up your patrol to seven men.' He shook my hand and walked back into the operations room – I was on cloud nine, not just for the promotion but also for the trust placed in me by him.

It's an old saying, but Rose was and still is a soldier's soldier. He would take never take crap, and was capable of holding an intellectual argument with senior officers and troopers alike. By far his greatest asset is his combination of a quick mind and swift action. These special skills have shown themselves on numerous occasions. Although 'pink listed' for the top, he came to prominence during the Iranian Embassy siege in London in 1980. After that, rumours abounded that he had become Prime Minister Thatcher's blue-eyed boy. Respect for Rose improved dramatically during the Falklands War. The trouble was, however, that he still liked to be in the thick of it, instead of being back at base watching the boys fly off on yet another mission.

Still, he did have the privilege of negotiating the ceasefire and surrender with the Argentines. There is a very strong rumour that Rose offered the Argentine commander in the Falklands, Major General Mario Menendez, a Harrier firepower demonstration as the Argentine troops massed around Stanley. As the surrender was agreed, there are rumours that he walked out of the building taking with him a prize statue of a horse. This statue had reputedly been given to Menendez by a grateful government for his capture of the Falklands.)

Rose's attitude to the Serbs was clear: back off or we hit you hard

Rose officially took over from Briquemont on 24 January 1994, two days after a mortar shell had killed six children in Sarajevo. His first task was to channel all information directly to himself so that he would not be operating in an information vacuum. He also made it clear that the aid convoys would no longer negotiate for passage. Days later, British soldiers fired back on the snipers who tried to prevent free passage of the convoys. His attitude to the Bosnian Serb leaders was clear: back off or we will hit you hard. It was a policy that would have worked, had it not been for the weakness of the UN and the indecisiveness of NATO. Agreement with the Serbs were littered with the latter's lies, deception and intimidation, tactics designed to win time and ground against the disadvantaged Moslems. These tactics angered Rose, and although he had employed the correct strategy in response, the West's politicians had not backed him, thus the key element of force was lost to him.

Towards the end of March the Serbs started shelling the Bosnian Moslems' eastern enclave of Gorazde; it was at this stage that Rose asked for SAS soldiers. Once it was realised that the Serbs were intent on capturing the city, Rose

Above: General Michael Rose, who commanded 22 SAS troops and also British forces in Bosnia as part of UNPROFOR.

tried by every means to arrange air cover. His main problem was the handful of British soldiers stationed there. The air support was not sanctioned, but a local ceasefire was arranged, which allowed the wounded to be evacuated, although two of them died. The only role to be found for the SAS was to act as observers and operate laser target markers. It is rumoured that two members of the Regiment were spotted and fired upon, as a result one was killed and the other badly wounded.

Farther east, the SAS helped save another society in 1991. With hindsight it was so obvious, so wickedly brilliant. There sat Kuwait, fat and ripe, bulging with enormous reserves of oil and cash, boasting an excellent port on the Persian Gulf and utterly incapable of defending itself against Iraq's proficient war machine. Saddam Hussein, the Iraqi dictator, was hungry for money but greedier still for regional dominance, and knew before the first of his soldiers crossed the border that it would be a walkover – it was. In 12 hours on 2 August 1990 Kuwait was his.

By 12 November the build-up of forces, under the flag of the UN, to recapture Kuwait was well under way. On the scorched sands of Saudi Arabia, 180,000 American ground troops waited impatiently, cleaning their weapons, exercising and thinking of D-Day. Flashing overhead were the best attack aircraft of the US

Air Force: McDonnell Douglas F-15 Eagle and General Dynamics F-16 Fighting Falcon multi-role fighters, and Lockheed F-117 Night Hawk 'stealth' aircraft. At sea, US Navy Aegis class cruisers programmed their Tomahawk cruise missiles to hit Iraqi targets, while aircraft carriers practised launch and recover operations with squadrons of interceptor and attack warplanes. Sitting on the sidelines in this huge military build up was Britain's SAS.

In the past few years, several books about SAS operations during the Gulf War have been written. Most have concentrated on the doomed patrol 'Bravo Two Zero', but in many other areas the SAS operated with great success. One of the first tasks undertaken by the Regiment was to rescue the hostages who had been seized by the Iraqis and used by Saddam Hussein as 'human shields'. Most of these were Westerners, although some were Japanese. Most had been seized during the invasion of Kuwait, and while the women and children were displayed

Below: One of the many SAS fighting columns in their Land Rovers behind the enemy lines in Iraq in January 1991.

before the world's media, many of the men were shipped out to important Iraqi military sites. The dispersal of the hostages made it almost impossible to rescue them (fortunately they were released before hostilities began).

The next task was to infiltrate deep into Iraqi territory and carry out search and destroy missions. The firepower carried was enough to take on and annihilate just about anything they could find, including the Scud launchers that were active in western Iraq. This mayhem would force the Iraqis to deploy large forces in order to locate them. A and D Squadron were divided up into mobile fighting columns, training for which had been carried out in the United Arab Emirates. It fell to the Mobility Troops to teach the rest of the boys the basic driving skills required in the desert. The hardest part was training the guys on the motorbikes. It is very difficult to control a bike in the desert, and it demands concentration and skill.

The Unimog can be loaded to the gunwales and still go anywhere

In each column a Unimog was used as the mother vehicle, which would carry the bulk of the stores. The Unimog's great advantage is that it can be loaded to the gunwales and still go anywhere. And they were loaded: rations, fuel, ammunition for several types of gun, NBC (Nuclear, Biological and Chemical) equipment and spares. The fighting vehicles in each column consisted of eight Land Rover 110s, each of which was armed with a Browning 0.5in heavy machine gun. Additional weapons included 7.62mm GPMGs, 40mm Mk 19 grenade launchers and Milan anti-tank missiles. The last was an excellent bit of kit for night operations: the thermal imaging sight provided good capability to a range of 8km (five miles) in total darkness, though the wire-guidance system limited the missile's range to 2000m (6555ft).

One of the SAS soldiers who took part takes up the story: 'Our original tasking took my column some 400km (250 miles) into Iraq, where we were to hunt in the southern central region, making our way up to the Euphrates. Morale was high: it was the type of role the SAS are well suited to and, providing basic SOPs were observed, contacts with the enemy would be on our terms and not theirs. It was a funny mood on the evening before we were deployed. This was real, we knew where we were going and that we were about to get ourselves into serious trouble. Friends who were allocated to different columns shook hands and made the

Right: One of A Squadron's Land Rover 110s deep behind the enemy lines in Iraq, February 1991, during the hunt for enemy Scud missiles.

odd joke, while others sat around and checked their personal equipment, or secreted their blood money about their person.'

(All members of the SAS were issued with approximately £1200 of blood money – in gold coins – and a blood chit written in English, Arabic and Farsi that promised the sum of £5000 to anyone aiding a British soldier. Each blood chit carried a serial number that could be checked against the person's name.)

'As with any SAS operation, the amount of personal equipment was vast. Going on exercise is one thing, but when you play for real you need the firepower. I personally carried an M16 assault rifle fitted with a 40mm M203 grenade launcher. For personal protection I also carried a Browning 9mm pistol. Back-up consisted of about 15 30-round magazines and a dozen grenades for the launcher. Tucked away on the back of my belt kit was a good medical pack and my survival kit, should escape and evasion become a reality. Our initial briefing was a load of absolute crap. Intelligence had said it would be warm, and accordingly many of the guys

were ill-prepared for the hazardous winter conditions we encountered.

'The Intelligence on the enemy was also very scanty. It was known that the Iraqi Republican Guard was well equipped and trained to a high standard. Yet there was very little information about the rest of the army or their locations – so much for all the technology spent on spy satellites! Not that it made much difference. Once behind enemy lines, no matter what we met, we were going to waste it.

'The bikes rode on ahead, or swept out to the flanks'

'The crossing for our column was made at a Saudi fort. From this location we could observe the border which, at this stage, was one big minefield. After observing the enemy for a day or so it was obvious that they did not move around much, but seemed to spend most of their time in the trenches. When at last the boss gave the go-ahead, we drove down from the Saudi fort and crossed the border – it was as simple as that. We drove in line formation and eventually picked up a track. The Unimog was positioned in the middle, as we couldn't afford the risk of it being hit. The bikes rode on ahead, or swept out to the flanks. This prevented us from being surprised, and also allowed us time to get the guns into line should the enemy suddenly appear. The bikes were also good for locating tracks and keeping us on route. If we were not sure, or the ground ahead could possibly conceal an ambush, the bikes went in first. As I have previously stated, riding the bikes was hard work, and they were constantly on the go.

It took no time at all for a pattern to emerge, and we soon settled down to our task. Some of the bikes would be used as scouts, while others would pass messages along the column. This negated the need for radio transmissions, while allowing the vehicles to stay in contact (the column could spread out over a kilometre). The lead vehicle always had a thermal imager that allowed the column to pick up problems in advance. The terrain in western Iraq was primarily lava bed, interlaced with very deep wadis. Getting into these proved a bit tricky, though luckily there were no ambushes.

'You really needed to grit your teeth against the weather. Here we were, dressed in tropical lightweight clothing, driving in open vehicles, and it was snowing. Thanks to the duff intelligence, most of the lads half-froze to death. But we were all in the same boat, and had to make do. As I drove the lead vehicle, my hands got so cold that they began to crack. We had been issued gloves but they were next to useless, and in the end I used socks as mittens.

'The first five days were relatively quite, which was just as well as it gave us time to settle down. During the day we rested in lying-up positions (LUPs), which we chose with care. We "pinged" (spotted) several enemy patrols, but our locations were well camouflaged and none came near us. The LUP would normally be in a depression, with several bug-out routes should we need to run, while any nearby high ground was used by our sentries. A great deal of care was taken with the vehicle camouflage nets, under which we would all hide during daylight hours. The daily routine consisted of posting sentries, cleaning weapons, sending sitreps (situation reports) and sleeping.

'An Iraqi officer, complete with map case and chart, climbed out of the truck'

'As mentioned earlier, the vehicles were well camouflaged during the day. We were therefore surprised when one of our patrols was visited by a Russian GAZ-69 truck full of Iraqis. They had obviously taken them for friendly forces, being so close to Baghdad. The driver stopped about 30m [100ft] away, got out and checked under the bonnet. An Iraqi officer, complete with map case and charts, then climbed out of the passenger side and walked towards the SAS position. All hell broke loose. After the shooting had stopped, all but one of the Iraqis were dead. It turned out that they had been a recce unit for an Iraqi artillery brigade, which was attached to a unit some 30,000 strong.

Left: Typical SAS dress for the appalling weather conditions in Iraq in early 1991. Note the well used M16 rifle.

Right: SAS fighting columns in Iraq during the Gulf War were frequently laid up throughout the day under camouflage netting.

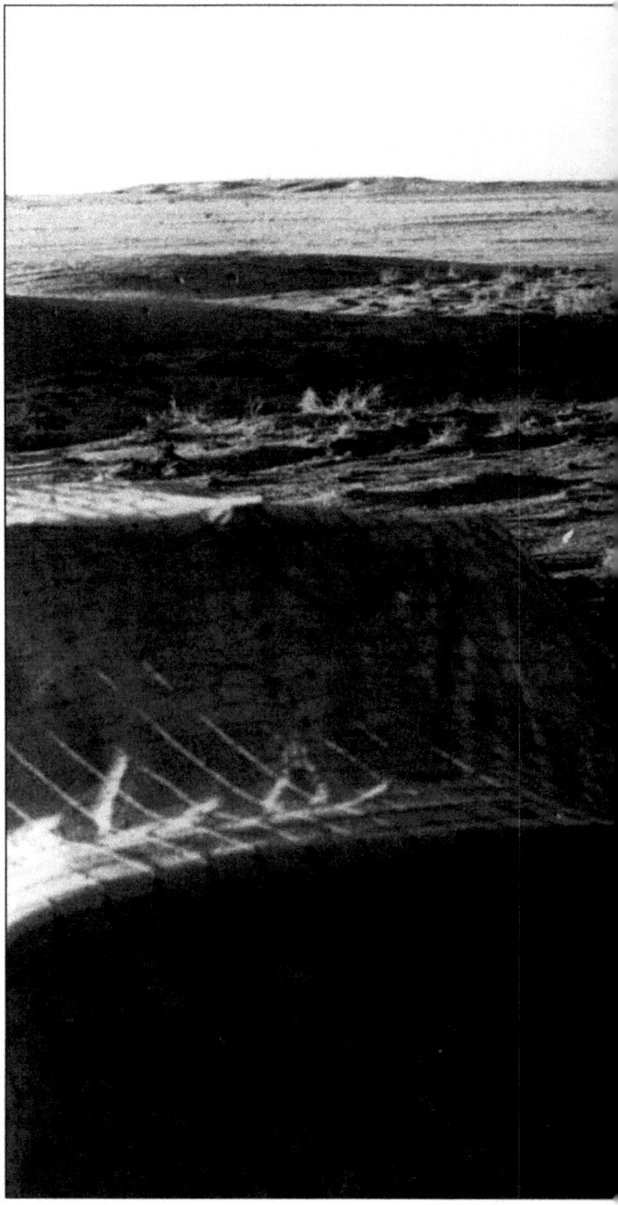

There is a joke in the Regiment that describes the incident. When asked where this vast Iraqi force was, the prisoner said "There," and pointed to the west. Two minutes later the SAS column was heading east like a bat out of hell. The prisoner was of great value, as were the plans and maps found in the vehicle, and it was decided to extract them both to Saudi Arabia. This was easier said than done, and required the whole column driving back through enemy territory to rendezvous with a chopper. All night they hit one Iraqi unit after another. Luckily for them, the Iraqis assumed they were friendly forces.

'After a few days the missiles started falling on Israel, and all SAS fighting columns were ordered farther west into the "Scud zone". Chasing the Scuds was a riot, and we had no shortage of targets. We would normally take a fix on the position and call for air support. Communications were not too good, as many of the radios we had been given were all set to different frequencies, and were not compatible for voice messages.

'While the demolition team went in, the rest of us kept watch'

'During this phase the head-shed sent us new orders. They wanted an attack on a microwave communications station. Our close target recce soon discovered the location. It was big – almost a kilometre [3300ft] square – with a vast tower covered with communication dishes. Time to plan our attack.

'As it turned out the attack went well. We managed to drive into the camp and park no more than 200m [660ft] from the main building and tower. While the demolition team went in, the rest of us kept watch. The demolition team had just completed its task and were withdrawing, when some Iraqi soldiers in a truck woke up and discovered us. There was no choice – we had to waste them. A firefight erupted. Luckily for us most of the dozing Iraqis thought it was an air attack, and so the majority of their fire went skyward. As the explosives dropped the tower, we all made a run for the vehicles. The team leader, in the front Land Rover, got a Global Positioning System fix and pointed in that direction. "Punch me a hole that way." Seconds later, several heavy machine guns opened fire and a lane was cleared. For an hour we ran the gauntlet of Iraqi soldiers, before disappearing into the empty darkness of the desert.'

As a result of their actions behind Iraqi lines, many SAS soldiers were decorated. A Squadron, for example, received a Distinguished Service Order, two Military Crosses, four Military Medals and four Mentioned in Despatches. Its losses amounted to two killed and one captured.

These are the big stories of rescuing society, most of them involving armed conflict. SAS soldiers, past and present, continue to make headlines wherever they are. One such case was the aid worker Don Reid from Hereford. Like so many before him, he had served his time in the SAS and gone off to find some new adventure. For him, it was working in war-torn Rwanda for the aid charity Assist UK. At the time he was driving a truck in an aid convoy, on the borders of Rwanda and Zaire. Unfortunately the area was full of rebels, many of whom displayed the normal tendency to shoot first and ask questions later. When his convoy came under fire, Reid thought it better to disappear. He fled into the African bush, whereupon he went directly into survival mode. The training had never left him, and by his own account he swam rivers, crawled through jungle and ate berries for food. All the time he kept sight of the Tonga mountains, where he knew there was a charity aid post. Eventually he found a track which led him to the Oxfam house. The news of his safe arrival was then passed to his own aid unit.

It was a minor incident, but one that gripped the imagination of the public. One man, a former SAS soldier, in hostile territory, pitting his wits against nature.

For the SAS, good relations with the local inhabitants are vital

Much as been written about the SAS 'hearts and minds' philosophy, yet little has been clearly explained. It is basically a logical way of befriending the indigenous people of an area in which the SAS is operating. Throughout history foreign armies have passed through or occupied towns and villages, normally exploiting and violating the local population. In the past this was done to feed the troops, satisfy the soldiers' sexual desires and gain information. Civilians were seen as an extension of the enemy's armed forces and, under this falsehood, treated accordingly. Whole populations have been massacred or forced into slavery as a result, and in more modern times many were sent to concentration camps. Recently we have had the horrific plight of the people of Bosnia-Herzegovina, and the scourge of ethnic cleansing. The result of all these actions has been to alienate indigenous peoples from foreign soldiers on their soil.

For the SAS, which operates behind the lines and in small groups, good relations with the local inhabitants are vital. One of the first chapters in SAS history in which locals have been helpful comes from an operation during the latter part of World War II.

Right: Any movement during daylight hours behind Iraqi lines was carried out at top speed to lessen the risk of being caught in the open.

Moussey is a very small town lying in the valley near the Vosges mountains of eastern France. To be more precise, it is 65km (40 miles) southeast of Nancy and 16km (10 miles) north of St Die. The village itself is spread out for about 1.5km (one mile), with odd houses dotting the roadside. Towards the centre is a church, beside which is a military graveyard. Many of those who lie there are from 2 SAS.

The unit had parachuted into France during September 1944, landing north of Moussey near Baccarat. It was to be one of the last

airborne drops the Regiment undertook during World War II. Soon after the drop the group moved to the woods and forest close to Moussey. The local population befriended the unit which, as time went on, provided them with some of the necessities of life and helped in whatever way its men could. The advancing US 3rd Army, under Lieutenant-General George Patton, was at that time temporarily held up due to lack of supplies. It was a delay the Germans took advantage of, moving reinforcements along the River Meurthe a few kilometres west of Moussey. The SAS men undertook numerous raids against the Germans, and in the end this brought retaliation. The Germans rounded up the population of Moussey and interrogated them. Not a word escaped their lips, and in frustration the Germans sent most of the men to concentration camps. Of the 210 men transported to those horrendous camps, only 70 returned alive after the end of the war. One tenth of the population was killed. Despite this, the SAS headquarters hiding in the nearby woods was never discovered.

When the SAS was reformed and sent out to Malaya in the early 1950s, it adopted a policy that is now known as 'hearts and minds', turning the concept into a weapon that could be effectively used against the guerrillas of the Malayan Races Liberation Army. General Sir Gerald Templer is credited with having first used the phrase 'hearts and minds', and he believed that it was possible to gain a moral victory over the enemy in a number of ways.

The first thing the SAS men did was provide a safe haven. In many guerrilla wars and other irregular conflicts, the local population finds itself trapped between the government forces and the rebels. In various remote regions, such as the jungle, the indigenous people are treated with brutality by each side. Guerrilla bands expect to be fed and accommodated and have their weapons and ammunition stored within the confines of the

village. When government forces arrive, the local population is accused of collaboration and slaughtered, a typical 'Catch 22' situation.

In Malaya in the 1950s the SAS put the 'hearts and minds' theory to the test. Jungle forts were built in aboriginal areas. This allowed the army to police the area and keep it safe from bands of Communist Terrorists (CT) as the guerrillas were known. In such a safe environment the locals could live, hunt and farm without being bothered. The soldiers not only protected the area but also established a medical centre, and above all the Regiment insisted that one man in each patrol speak the local language as this led to better understanding and the transmission of information.

More than half the squadron contracted one disease or another

In 1953 Fort Brooke was founded. A patrol of D Squadron had predetermined a location from the air and, after a long, hard trek through thick jungle, finally arrived at the site. The SAS stayed in the location for three months, during which they built a helipad, constructed a bridge and made friends with the locals. Once the base had been established, SAS patrols cleared the area of any CT camps. This all sounds very easy, but it was a costly exercise. The Regiment lost two men in a CT ambush, and more than half the squadron contracted one disease or another and had to be evacuated. The SAS men got the job done, but it was a ragged and emaciated SAS squadron that handed over the location to the local police.

This policy of going into the enemy's back yard and establishing a friendly base has worked on many occasions, and in Oman it was implemented on the Jebel Dhofar. This mountain massif forms a huge natural barrier at the edge of the inhospitable desert. It is a giant slab of rock extending inland from the Salalah plain, running parallel to the coast and stretching for 240km (150 miles) right down to the border with Aden. As mountain ranges go it is not a particularly high range, reaching only 900m (1985ft) or so at its highest point. It rises from the plain to a plateau which narrows to the east and west, but is some 15km (nine miles) wide in the centre.

In the extreme east you can find scant, barren soil and some limestone, all broken up with sparse vegetation and little or no water. But as you travel westwards, patches of rich soil begin to appear, interspersed with clumps of scrub-land grass. In the centre of this region the soil suddenly becomes deeper, covering the plateau with a lush carpet of green grass and stunted trees, while large valleys, known locally as wadis, run southwards to the sea. In some of the larger wadis are rivers and streams, providing sustenance for lush vegetation and forming natural water holes in the rock structures. Here the land is capable of sustaining human and animal populations alike. It was here, among the local population, the *adoo* (see Rescuing Our Comrades, pp80-99), that the enemy established their base.

'Some of G Squadron set off on a "hearts and minds" campaign'

The author takes up the story: 'Once the war was taken onto the jebel, many of the *firqat* (see Rescuing Our Comrades, pp80-99) visited old friends and relatives that lived among the Jebali hill people. This was encouraged by the SAS as it normally resulted in the *firqat* returning with good information. I often looked at the local Jebalis and thought in my arrogance "How can they live like that?" Yet the question is not mine to ask. They have lived here, building the same stone huts, raising children and providing for their families long before Western man had learnt to read or write – and they have survived. They are a simple people bound by unwritten laws, and war with all its horror was but one of their burdens.

'After the abortive raid on the northern coast of Oman and the freefall parachute descent into Wadi Rawdah (see Rescuing Our Allies, pp30-53) some of G Squadron, including myself, set off on a "hearts and minds"

Left: Fort Brooke, Malaya, one of the many jungle forts established by the SAS during the Malayan Emergency.

campaign. The trip was to take us by Arab *dhow* (small sailing ship) around the Straits of Hormuz, where we would visit the tiny coastal villages of northern Oman. The *dhow* was crewed by three Arab soldiers from the Trucial Oman Scouts and three Royal Marines of the SBS, and a Rigid Raider craft also accompanied us. As many of the smaller villages had no docks, the Raider was used to put us ashore (it was also used for water-skiing). The journey and life onboard were quite relaxing: we slept on the deck, washed each morning by jumping overboard and fed on fish from the sea.

'We were the first white visitors they had ever received on the island'

'It was not until we reached our first village that the term "hearts and minds" really came home to me. I say village, but it was nothing more than a few stone huts and half a dozen people. A party of eight SAS went ashore, two of them medics, of which I was one. I have seen poverty, but compared with this place Western poverty was the Ritz. The family had apparently committed some crime in the past and had been banished by the Sultan to this remote and barren island. No one could remember the crime, or how long the family had been there, but it was several generations ago. They had one small boat, more a canoe really, fashioned from a dead tree trunk. The fish they caught, plus a little rice, was all the food available. A well had been dug at the farthest point from the shore, but due to the rocky features of this desolate place it was no more than 50m (165ft) from the shore. I tested the water – it was so salty that none of us could drink it.

'The head of the village, an old man who looked 80 but was probably about 40, welcomed us to his home. We were the first white visitors they had ever received on the island, and he celebrated with typical Arab hospitality: and the best food in the house was offered. This turned out to be a 20-year-old can of pineapple. The label had long gone and the contents had turned to mush. We sat there in our clean combat uniforms, healthy and well fed. To a man we were all humbled by the old man's gesture and ate the fruit. I later examined one of the women who was complaining of toothache. She looked old, but gave her age as 25. I could not believe the state of her teeth, for many were missing and her gums were a rotting mess of blood and puss. Gripping one of her teeth with a pair of forceps, I was dismayed to discover that it was very loose – a gentle tug and it came out. In the end I took out all her teeth and did my best to sterilise her mouth. We remained on the island for three days, and I continued to treat the woman. I have always been amazed by the speed at which penicillin injections worked on these people. Before we left for our next destination, all the ration packs were stripped of jam, butter and biscuits and presented to the old man, with spare clothing going to the children. A signal was also sent requesting that the family be repatriated to the mainland immediately.

'In the "hearts and minds" campaign medics have played a very important role, so it is imperative for the Regiment to train them well. The initial purpose of the SAS medic is, of course, to treat the members of the patrol of which he is part. The standard of training required is very similar to that of a paramedic in civilian life. His role is to keep his patient alive until more professional help is made available, but in some situations he is the best and only medical support available.

'The SAS medic removed the festering skin and the skull bone'

'In a somewhat larger coastal village in Oman, a worker had fallen off the top of a building. His fall had resulted in a large square piece of his skull becoming detached. The SAS medic removed the festering skin and the skull bone from around the wound; the brain was clearly visible. He next requested that the lid be taken off the top of a large tin of beans from a ration pack, which was then beaten flat and placed in boiling water. I laughed when he said he was going to use this to cover the open wound. But he made an excellent job of it, pulling the clean skin over his handy work and suturing it

Above: Another of the jungle forts set up by the SAS and then garrisoned by regular British Army infantry and artillery.

securely. The wound was treated with penicillin powder and he was left with further medication. Several months later I returned to the village. As I walked up the beach the man was paraded in front of me. Smiling, he bent his head and tapped the tin lid. The wound had healed extremely well, but a small section of the tin could still be seen, the letters "ens" from "beans" was still clearly visible.'

In many parts of the world water is the most precious commodity. For centuries isolated Arab villages have relied on digging small wells, and villagers would trek kilometres each day just to fill a few goat skins with brackish water. So the effect of a British well-drilling team can be quite phenomenal. As the SAS teams spread out over the Jebel Dhofar and northwards into the more desert areas, they brought with them a drill team. Within days fresh, clean water would fountain up and flood the desert. To the local population this was little short of a miracle.

During the monsoon season, when helicopters could not reach the locations due to low cloud, the delivery of fresh food was limited. Living off 'compo' rations is far from

Right: Dhofaris with leaflets dropped by the Omani government following the amnesty in 1970, inviting Dhofaris to join the government.

ideal, and at one location "Jebel Bakeries", as it came to be known, was born. It started with one SAS man balancing an empty mortar box on top of two Primus stoves to create an oven. Trading with the *firqat* for bags of flour, the first bread was baked in a mess tin. At first this rare luxury was confined to the SAS group only, but as news of the bakery spread the demand grew and people would walk kilometres to get their daily loaf. In the end the bakery was going non-stop from dawn to dusk.

During the months that followed the monsoon, the lush richness of the central massif was capable of growing food. The SAS teams led by example and often cultivated gardens to grow fresh vegetables.

'Hearts and minds' is also concerned with the enemy

In the 'hearts and minds' war language is also important. Communication brings understanding, which in turn breeds trust. For the SAS most language courses start off at the Army School of Languages in Beaconsfield, although some are also taught at Stirling Lines. Once the course is completed, the first priority is to get the soldiers into an environment where they can practise. One such soldier, a trooper of long standing, learnt Malay. At the end of his course he was sent off to Malaya to live with a local tribe for a few weeks. Normally in such circumstances the soldier would use his common sense and periodically make some form of contact with base. The reason for this particular SAS soldier's continued rank of trooper was that he had no common sense: he never actually did Selection as he was one of the ex-Guards Parachute Company men who slipped in when G Squadron was formed. No one noticed the guy was missing until his presence was required months later, and then a search party had to be flown to Malaya to locate him. They found him deep in the jungle living with a local tribe – he had turned native.

But 'hearts and minds' is not just about water, food and friendship. It is also concerned with the enemy. If some of the enemy can be persuaded to change sides, there is much to be gained. During the Oman War, leaflet drops were made in an effort to explain what the new Sultan was trying to achieve. As a result, many of the *adoo* left the communists and returned to the fold. Likewise, many *adoo* were taken prisoner, and these were also given the choice of

being shot, going to jail or joining the *firqat*. Those who joined the *firqat* were quickly put in a chopper and flown over *adoo* territory. All he had to do was point to his old hiding places. These flights were know as 'flying fingers', and the chopper was normally followed by several jet fighters, which then bombed the indicated area with 227kg (500lb) bombs.

When these skills are used in warfare within the context of a 'hearts and minds' effort, they help the men of the SAS to develop a brotherhood and kinship with the indigenous people. The basics of life in most war-torn areas are always difficult, and any effort to improve them is generally welcomed. The rewards, as General Templer so rightly pointed out, are shelter, trust and information. Sometimes, as in Moussey, the price is high, but once the trust has been established the bond is generally for life. And the rewards, as in Malaya and Oman, are great.

RESCUE IN NORTHERN IRELAND

In the legal and political minefield that was the undeclared war in Northern Ireland, SAS soldiers had to constantly watch each other's backs while fighting the IRA.

Northern Ireland was never an easy place for the Army, and in particular the SAS, to work in. Firstly, in purely military terms the battle against the Irish Republican Army (IRA) could have been won years ago; all the terrorist 'players' were, and are, known to the security forces. Secondly, units such as E4A (an undercover surveillance team made up from the Royal Ulster Constabulary Special Branch and trained in Hereford by the SAS) were far better adapted to sorting out their own problems, even if at times they were a little over-zealous. But Ulster was never a purely military affair; instead, Northern Ireland became the haven of the military intelligence officer. Gathering information, tasking Close Observation Platoons (COPs) and pretending to be in the 'know' were all one big game.

(Author's note: Even when you were confronted by a member of the IRA, complete with gun or bomb in hand, you had to ask him nicely, 'Stop, or I will shoot', and repeat

Left: A patrol waits to be collected after a reconnaissance mission in the 'bandit county' of South Armagh, an area of high Irish Republican Army (IRA) activity.

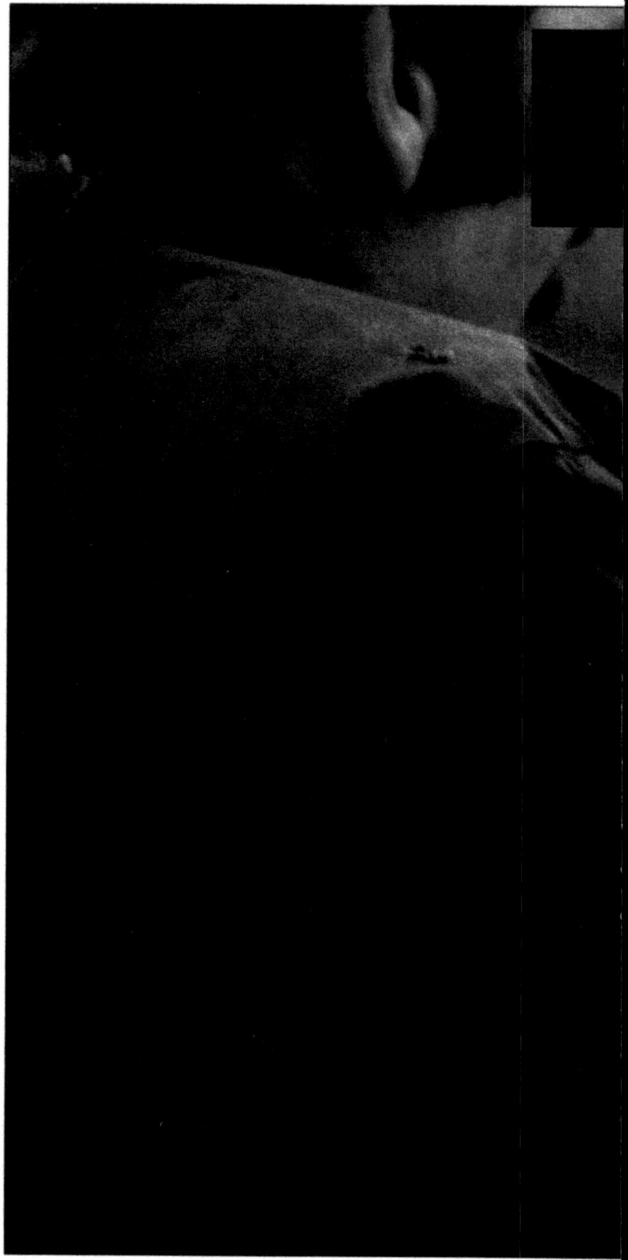

Right: A plain clothes SAS soldier in Northern Ireland with an Ingram SMG which was rejected for political reasons.

it three times. Then, if the suspect had not already killed you, and you managed to get a quick shot off, killing him, you were in real shit. Just pray that the suspect, in his moment of indecision, did not turn to run and that the bullet you put in him did not enter his back. Why? Because they are going to hang you. For some reason, British politicians wanted the soldiers in Northern Ireland to fight the IRA while walking on a tightrope – one slip and you had had it. In court, events could get twisted out of all recognition. A typical example is as follows:

'Did you take aim?'

'Yes sir.' *If you say no, you will be charged with being erratic.*

'So you hit him with the first round.' *Enticing you to say yes.*

'Yes sir.' *You make the big mistake of being proud of your marksmanship.*

'So how was it that there were four entry wounds to the suspect? And did you not fire seven rounds of ammunition?' *Talk your way out of that one.*

The 9mm Ingram Model 10 submachine gun is a neat little gun

'The most stupid thing is that once a shooting has taken place, all concerned are isolated until the 'legal eagle' (military lawyer) arrives. He then takes everyone's statement and formulates what happened. Everyone is told what to say in order to keep the story consistent, leaving no holes for the prosecution. If they find any inconsistencies, the case will be referred to the civil authorities and those involved will stand trial. It has happened several times to members of the SAS. Yet the author has never understood why, in a firefight with a terrorist, it matters if the enemy was shot with one round or several. But such was the situation in Ulster.

'Another prime example of political stupidity regarding operational procedures in Ulster was the occasion when the SAS wanted to use the 9mm Ingram Model 10 submachine gun. This is a neat little gun, and if required can be used single-handed. Both it and the old Sterling, which was being used at the time, fired exactly the same ammunition. It took a meeting with the top brass in Northern Ireland to convince them which was the better weapon. Their main worry was the political implications

of shooting a member of the IRA with an unorthodox weapon. They could not see that it was the same round regardless of the weapon. But this is the reality of life for the SAS in its fight against the IRA. No matter what happens, if the SAS is involved in a shooting then it is automatically accused by the IRA of committing murder, and this in a 'war' in which the IRA claims its soldiers are on 'active service'!'

Whatever the problems, though, the Regiment was firmly committed to Northern Ireland. Its members had been in the Province since 1973, but most of these were just visits, where an SAS man would be attached to a local regiment. Then, in 1976, the Regiment became a political tool, when the government announced that it was sending it to Northern Ireland to curb the increase in IRA violence.

Results came quickly: on 12 March 1976, Sean McKenna, a known IRA member, was lifted from his home over the border in Eire. He was dragged from his bed and frogmarched into Northern Ireland, whereupon he was handed over to the RUC. Less than a month later another IRA member, Peter Cleary, was lifted from the home of his fiancée, who lived just north of the border. The house had been under observation for some time as it was known that Cleary was soon to be married, and it was just a matter of waiting. Once the suspect had been taken into custody, the SAS men moved to a pick-up point and waited for their helicopter. However, Cleary tried to escape, and was shot dead as he did so. The incident did not go down well. The IRA claimed that he had been murdered.

Left: A British Army soldier on patrol in Belfast. Ulster was fraught with difficulties for both the Army and the SAS.

Several senior officers in Northern Ireland were aghast at having a bunch of men such as the SAS within their midst, but as the prime minister had sent them in there was little they could do. However, the situation did serve to send a message to the IRA: border or not, we will come and get you. This theory was confirmed when, at 18:00 hours on 5 May 1976, the Eire police stopped two men in a car at a checkpoint within the Republic, and in the process discovered that it had members of the SAS on its hands. To make matters worse, two back-up vehicles arrived to assist the first, and in total eight SAS men, together with their vehicles and weapons, were taken into custody by the Eire police. As the news emerged all hell broke loose, and the newspapers had a field day. The men were taken to Dundalk and then on to Dublin. All were charged, and in the end it was just as embarrassing for Eire as it was for the SAS soldiers, most of whom were fined £100 each for having unlicensed weapons.

'As well as several cut-off groups, a reaction team lay in wait'

During the late 1970s there were other successes and the odd failure, but slowly the Regiment evolved its own strategy, forming stronger links with the RUC's Special Branch. This link provided crisper information for the SAS to work upon. One such operation came in June 1978, when information was received that an IRA cell was about to bomb the Post Office depot in Belfast.

A member of the SAS who served in Ulster at the time takes up the story: 'It was a classic operation. Special Branch had given us the information and we would provide the solution. It was fairly simple: a team of IRA terrorists were to firebomb the Post Office depot in Belfast. We even knew of their approach route, and made our plans accordingly. The only thing we did not know was the exact time of the bombing. In the event several observation posts were set up, the main one being located in a house. This gave us a clear view of the small alley that ran by the side of the Post Office compound where the vehicles were kept. This compound was protected by a high fence.

'As well as several cut-off groups positioned around the compound, a reaction team, sitting in a van, lay in wait. I hate to think how many hours I spent lying in that van – it was bloody uncomfortable. As the operation progressed over several days, it was deemed a better idea to establish two guys in a large bush during the hours of darkness. On the night in question the task fell to Tony and Jim (not their real names). Tony was in my troop, and a better man is hard to find. He was solidly built, laid-back and very cool. In this incident and many others afterwards, when the chips were down, he was the man to have by your side. Jim was a little younger and not quite so level-headed.

Below: An SAS soldier in his Observation Post (OP) in South Armagh. The broken lines denote the levels of water he had to endure.

'The SAS guy in the observation house was bored. He had sat at the window for several days and nothing had happened. Special Branch had reassured him several times that the bombing would take place, but as time went on he had begun to entertain doubts. Then, suddenly, he saw some figures in the alley approaching the wall of the compound. Before he could raise the alarm one of the figures moved his arm, and the first bomb was already arching its way towards the target.

'Tony fired two short bursts, and two men fell dead'

'Orders were screamed: "Go, go, go!" The operation was activated. With Tony in the lead, the two men leapt from the bush. The light on top of Tony's MP5 submachine gun sent out a beam which illuminated three figures in front of him, two of them about to throw satchel bombs. Tony fired two short bursts, and two men fell dead. The third made a run for it, retreating back down the alley in the direction from which he had come. Tony fired another short burst and the third man fell. It was perfect except that Jim, who on this occasion was armed with an SLR rifle, decided to confirm the kills, pumping a lot of heavy calibre rounds into the bodies. Almost as Jim finished firing, two more men entered the alley from the bottom end. Tony immediately issued a challenge, unsure as to the identity of the men. One dropped to the ground and placed his hands on his head, but the other made a run for it. He never made it – a short burst from Tony killed him outright.

'As the situation stabilised, it turned out that the two men were just returning from the pub and had accidentally walked straight into the firefight. Had the man, a Protestant by the name of William Hanna, remained still after the challenge, he would be alive today. And I can assure you a clear challenge was given. Without going into too much detail, the fun really started when the legal eagles arrived. The weapons were tagged and taken away for forensic examination. Then each man gave his story. It was all clear and simple until the lawyer asked Jim to give his account in his own words. "Well, I saw Tony move and I backed him up. He slotted two, but the third made a run for it. As Tony fired, I stepped to one side and shouted to the boys in the bottom cut-off. 'Keep your head down, Bobby.' Then I let rip with a 30-round mag." The legal eagle looked in horror, while the rest of us just rolled around laughing. Subsequently his story was corrected, and no court case followed.

'Not long after, Special Branch in Armagh tasked the SAS unit in Portadown to undertake a reconnaissance of a small Catholic church in Castleford. The essence of the information was that there was an IRA meeting place in a small building at the rear of the church grounds. Special Branch requested that the SAS take a look, and photograph any documents of interest that was discovered.

'One of the SAS men produced a set of picks and made short work of the lock'

'Two men were entrusted with the task, one of them a lock-picking specialist. They were driven to the town by two other SAS personnel, who would stay mobile in the area and act as back-up. At around 01:00 hours on a wet and windy morning, the two men were dropped off by a gateway at the edge of the town, and quickly made their way to the rear of the church. The whole church was surrounded on three sides by a 3.65m (12ft) wall, with an iron railing fronting the road. Approaching from the rear, the two men quickly climbed the wall and, remaining in the shadow of the church, approached the meeting house. The small, two-storey building stood in one corner, with a single door providing the entrance.

'One of the SAS men produced a set of picks and made short work of the padlock. Once inside they searched the ground floor. There was nothing of interest as most of the space was taken up with church props and junk. A small set of stairs led to the second

Right: SAS soldiers taking the chance to relax and enjoy themselves in the mess at Bessbrook base in South Armagh.

floor, the entrance to which was also barred by a padlock (which was soon picked). This upper floor had two windows, allowing the single room to be illuminated by the street lighting. The room itself contained several lockers filled with paperwork and books, with a large table and several chairs sitting in the middle.

'The two men carried out a search of the room and the cupboards, but little of interest was found, and certainly nothing referring to the IRA. A small trap door leading to the loft was discovered, and by placing a chair on the table it could be reached. Checking the hatch for any booby trap, one of the SAS men climbed inside the loft. After he had secured the hatch behind him, he switched on his torch. What confronted him was a whole pile of large plastic bags, most of which contained bomb-making equipment. Several pounds of Semtex plastic explosive were also discovered, as were 20 ready-to-go cassette incendiaries. The information was relayed directly back to control and an observation unit organised for the next day. Unfortunately, when Special Branch found out about it the situation changed. For some unexplained reason, MI5 became involved; this

meant no SAS allowed anywhere near and the target being electronically tagged. The two SAS men had the pleasure of taking the MI5 operative into the building, but he was so fat that he demolished half the rear wall as he climbed over it! Then he could not pick the lock, and thus became hell-bent on calling in an expert, until the SAS man picked it.

'For the SAS that was the end of the story, although only for three weeks. Then a call came from Special Branch to remove the whole explosives find and bury it in a hide. The hide was to look like an IRA job and be at least half a mile from the church. This we did, and next day an SAS soldier stood on the site as the local Ulster Defence Regiment (UDR) unit searched

Left: South Armagh. Many IRA terrorists roam back and forth across the border at will – a tactic also used on occasion by SAS patrols.

the area. "It's here mate," said the SAS man. The day after, the newspapers carried headlines of a major explosives find by the UDR. There was a reason for this subterfuge: it told the IRA, which had put the explosives in the church grounds, that it was being watched, and in the eyes of the general public it also made the British Army look good.'

There was also a time in Northern Ireland when nothing seemed to go right. April 1980 was a perfect example. The SAS in South Armagh had been tasked with several jobs all at the same time. The first was to set up an observation post (OP) watching the home of two RUC men whose lives had been threatened. The second was to keep watch on a public house the IRA intended to firebomb. Finally, a well known IRA terrorist intended to shoot a man suspected of being an informer at a dance. The task of the SAS was to keep tabs on both the gunman and the target.

As the two RUC policemen returned home, their car was shot up

These three operations called for more SAS men than were available, and so more troopers were drafted into South Armagh. Four men went into the first OP watching the two RUC policemen. Four more went into the OP at the pub, but numbers were still short, and so this OP was pulled off during the operation against the would-be IRA assassin and his target. The latter required at least six cars, each carrying two men, plus several men in the dance hall or pub. In addition to all this, men were needed for drop-offs, re-supply, the Quick Reaction Force (QRF) – heliborne troops providing instant back-up – and to man the operations room.

Things started going wrong when, after a week, the OP watching the RUC policemen was compromised by a farmer and asked to be extracted. Special Branch was informed of this and pleaded with the SAS team to stay put. The commander of the OP made a decision and, as with all SAS operations the man on the ground calls the shots, the men were withdrawn. Next morning, as the two RUC policemen returned home from work, their car was shot up (spent cartridge cases were found just a few metres from where the SAS's OP had been). Luckily the two policemen were unharmed. A week later, as the OP on the public house was

Left: Notwithstanding the 1994 ceasefire in Northern Ireland, the IRA remains a well armed and ruthless terrorist organisation.

withdrawn to help with the surveillance on the would-be assassin, the second disaster happened. By 02:00 hours on Sunday morning, both the IRA gunman and his target had been followed. The target had reached his home without incident. As the cars drove back to Portadown Barracks, their route home took them past the pub that had been under observation – it was burning.

By this stage Special Branch was displeased with the SAS, but still the operation continued. In Belfast it was discovered that the terrorist weapons under observation had been moved. Unfortunately their new home was a bit of a mystery, and the nearest location that could be found was a block of three houses in a terraced row on the Antrim Road. After a check with Intelligence, it turned out that one of the three houses had been previously used by the IRA, so this house was targeted. On the afternoon of 2 May, two cars headed down the Antrim Road and screeched to a halt outside No 369, another vehicle containing three SAS men securing the rear. For security reasons there had been no cordon or military activity before the raid, and the boys steamed straight in.

The IRA had mounted a 7.62mm M60 machine gun in the upstairs window

The lead groups were already in the house and up the stairs, when a burst of automatic gunfire filled the air. Unknown to the SAS team, the IRA had mounted a 7.62mm M60 machine gun in the upstairs window of the adjoining house. The boys had moved in fast, but the commander, Captain Richard Westmacott, had been sitting in the middle of the rear seat and so was the last to move. He was shot dead by a burst from the M60. Realising what was happening, the whole assault was quickly switched, but by this time the IRA man had surrendered. The sound of gunfire brought the Army and RUC to the scene, and a Catholic priest materialised to see that the IRA man was allowed to surrender. Other suspects leaving by the back door were arrested, and weapons were also recovered.

Captain Westmacott had joined the SAS from the Grenadier Guards, and was an officer in G Squadron. He did not personify the typical SAS officer, for his fair curly hair gave him the look of a schoolboy and he had a love for poetry, which he had learnt as a child sitting on his grandfather's knee. Inwardly he was as tough as they come, and he rightly won a posthumous Military Cross. His death ended the run of bad luck that had dogged the SAS in Ulster, and the fortunes of the Regiment were further improved when the stormed the Iranian Embassy (see Hostage-Rescue, pp54-79).

Despite being hit several times, the driver managed to keep going

More than two squadrons were committed to the Falklands War, so for a short while Northern Ireland took a back seat in the Regiment's priorities. But by December 1983 Northern Ireland was back with a vengeance. It was known that a part-time RUC officer was about to be shot, so the target was identified and all efforts were made to provide protection. At the same time the IRA weapons intended for use in the assassination were located. They were discovered under a large thorn bush in a field near Coalisland in County Tyrone. A small country lane ran past the field, which was separated from the road by a thick hedgerow, but with a gateway providing easy access. A full-scale operation was mounted that involved No 14 Intelligence and Surveillance Group, which would keep tabs on the IRA players; the SAS, which would act as cover for the weapons; and E4A, which would form the QRF. Additionally, a platoon from the Devon and Dorsets would be called upon to supply a cordon once any incident had taken place (this is a very strong pro-IRA area, and any Army activity normally draws a hostile crowd).

The SAS inserted a two-man OP in a ditch opposite the weapons hide, some 20m (66ft) away. Two more OPs were inserted, one at each end of the road, to act as early warning and vehicle stops. E4A was positioned at

Lurgan. After several days of waiting, news filtered through that the hit would take place on the Sunday evening. For the SAS men in the OPs this could not come soon enough – the December weather was wet and cold. The men could only eat and drink cold food, and re-supply was not allowed for fear that tell-tale signs would be left behind. At approximately 15:30 hours in the afternoon, one of the stops reported a car approaching. The vehicle stopped opposite the gate. Two men got out and climbed the gate, leaving the driver in the car. The men then made their way directly to the weapons. Kneeling down to retrieve them, the first terrorist, Colm McGirr, pulled out a weapon and passed it to Brian Campbell. As Campbell turned back in the direction of the car, the two SAS men stood up in the OP and made a challenge. McGirr, who was still kneeling by the bush, turned with a gun in his hand and was shot dead. Campbell ran for the gate, still holding his weapon. Two more shots rang out and he fell mortally wounded. On hearing the gunfire, and realising that it was a trap, the third man in the car drove off. The SAS stop jumped out to halt him, firing four high-velocity rounds directly into the vehicle. Despite being hit several times, once in the head, the driver managed to keep going, making his way to a small housing estate about 3km (two miles) away. E4A gave chase and the car was quickly located, but unfortunately the driver had been whisked away by some of the many local Republican sympathisers. He turned

Left: The date was 8 May 1987. The place was Loughall in Northern Ireland – the scene of one of the SAS's greatest victories.

up some hours later in a hospital over the boarder in Eire.

Once the area was secure, Campbell was given emergency medical treatment by an SAS medic at the scene. The wound to his lung was dressed and an air-way was inserted, but it was too late and he died from his injuries about 10 minutes later. There was the usual outcry from the IRA and its sympathisers, and a little bit of bad press. Most of this was based around a report from an ambulance driver who had visited the scene. In the eyes of the SAS, though, it was a clean, neat job, apart from the vehicle stop. The SAS soldier was severely criticised by the Regimental head-shed for allowing the car to pass, and the man was later RTU'd (returned to his original unit).

Another classic SAS operation took part at Loughall on Friday 8 May 1987. Intelligence had been received to indicate that the police station at Loughall was to be attacked by the IRA, using the same method as it had employed a year before in County Armagh. That incident had taken place in April 1986, when a mechanical digger had been packed with explosives and driven into the RUC station at the Birches, causing widespread damage. A report that another JCB had been stolen in East Tyrone gave rise to the belief that an identical IRA operation was being planned. All efforts were made to locate the digger and identify the target. After an intense covert search, the weapons and explosives were located. Subsequently, the digger was also located in a derelict building on a farm some 15km (nine miles) away. Surveillance by E4A provided more information, and eventually the target was assessed as being the RUC station at Loughall. This station was manned on only a part-time basis, and consisted of one principal building running parallel to the main road and surrounded by a high wire fence. The time and date of the attack were eventually confirmed through a Special Branch telephone tap.

At this stage the SAS party took up its ambush positions

Two of the IRA activists were named as Patrick Kelly and Jim Lynagh, who commanded the East Tyrone active service unit. When masked men stole a Toyota van from Dungannon, Lynagh was spotted in the town, and this suggested that the van might also be used in the Loughall attack. Not long after this, the OP reported that the JCB was being moved from the farm. At this stage the SAS party, which had been reinforced from Hereford, took up its ambush positions. It was later reported that some of the SAS men were in the police station itself, but this is not true. Instead, most of the main ambush party were hiding in a row of

Right: The bullet-ridden IRA van after the Loughall ambush. Eight terrorists died during the action, which lasted only seconds.

small fir trees lining the fence on the opposite side of the road to the station. Several heavily armed stops were also in position covering all avenues of escape.

At a little past 19:00 hours, the blue Toyota van drove down the road in front of the police station, and several people were seen inside it. A few minutes later it returned from the direction of Portadown, this time followed by the JCB, in whose cab where three hooded IRA terrorists. Declan Arthurs was driving, with Michael Gormley and Gerald O'Callaghan riding shotgun. The bucket was filled with explosives contained in an oil drum, which had been partly concealed with rubble. While the blue van charged past the station, the JCB slammed through the gate. One of the two terrorists riding shotgun (it was not very clear which one) ignited the bomb and all three men made a run for it. Back at the van, several hooded men jumped clear and started to open fire in the direction of the RUC station. At this stage the SAS ambush was activated.

Loughall was one of the most successful operations mounted against the IRA

The sudden hail of SAS fire was terrifying, and all eight members of the IRA were cut down. At the height of the firefight the bomb exploded, taking with it half the RUC station and scattering debris over all concerned. As the dust settled, the SAS closed in on the bodies. At that moment a white car entered the ambush area, and both its occupants, who were dressed in blue boiler suits similar to those worn by the IRA terrorists, were unfortunately mistaken for more terrorists. It did not help the cause of the two in the car that, on seeing the ambush in progress, they stopped and started to reverse. One of the SAS stops opened fire, killing one of the occupants and wounding the other. It later transpired that the dead motorist, Antony Hughes, had nothing to do with the IRA. Several other vehicles and pedestrians soon appeared on the scene, but by this time the situation had been stabilised. Loughall was an incredible sight: the RUC station half demolished, the mangled yellow metal that had once been a JCB, and the numerous bullet-ridden bodies littering the road.

Without doubt Loughall was one of the most successful operations ever mounted against the IRA, which was totally stunned by the loss of two complete active service cells. The Hughes family were compensated for their loss, and with no public inquest the matter was closed. The IRA, believing that there was a mole in its organisation, went into a period of self-assessment. For the SAS, Loughall was a victory, but the war went on. And the Regiment maintains its commitment to rescue Ulster's citizens from the scourge of terrorism.

RESCUING VIPs

When VIPs need safeguarding, both at home and abroad, the British government assigns SAS soldiers to get the job done properly. Their bodyguarding skills are the best in the world.

The SAS has often rescued VIPs in various parts of the world, but it was in Aden that the Regiment's bodyguarding skills were first refined. Once a bustling, commercial city which had been under British control since 1839, the port of Aden had been developed into an important refuelling facility for shipping operating through the Suez Canal. It had also become a vital military installation because of its strategic position on the sea route between Europe and the Far East. By 1967, though, the streets, once packed with tourists, were patrolled by British soldiers in an effort to stem the violence leading up to the British withdrawal in November of that year. The unrest was orchestrated by the terrorists of the National Liberation Front (NLF) and the Front for the Liberation of Occupied South Yemen (FLOSY).

As Aden closed down, the British Army started its withdrawal. This was not just a matter of pulling out the military units scattered around the city itself, but also entailed gathering up the outlying stations and individuals. Around this time, information came in that a large enemy force was about to attack an important RAF base. The base was isolated, and so G Squadron was sent to protect it. The

Left: Members of the British Royal Protection Group undergoing training under the ever-watchful eyes of the SAS at Hereford.

Right: Members of the Italian hostage-rescue unit *Nucleo Operatives Central di Sicurezzo* (NOCS), which was established by the SAS.

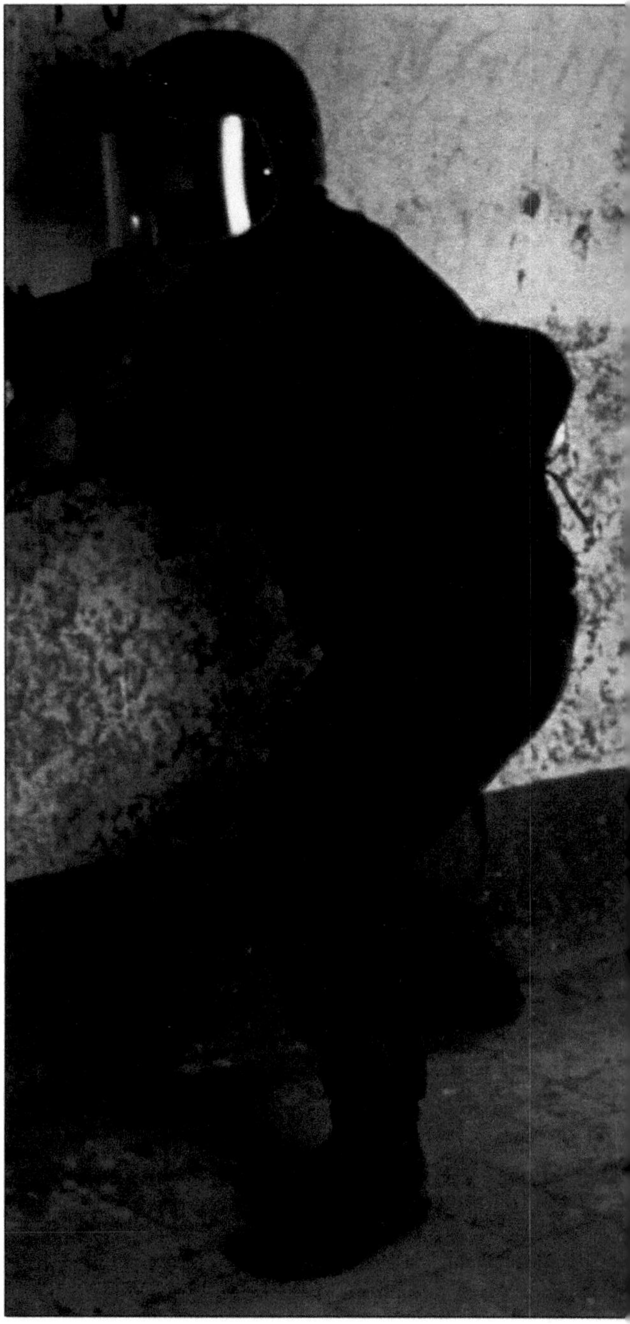

camp was fairly large, and had been used by the RAF as a refuelling station for aircraft travelling between Europe and the Far East. As the camp, containing 200 or so inhabitants, made ready for the withdrawal, the SAS carried out reconnaissance in the area. Several long-range patrols were undertaken by Mobility Troop, but there was no sign of the enemy.

Some two weeks into the operation, at 16:00 hours in the afternoon, a 'Flash' signal was received by the base commander. The camp was to be abandoned. All documents were to be destroyed and only personnel would be extracted (though the camp itself was not to be destroyed). The SAS had received separate orders to extract several VIPs from three separate locations. This was a simple task: two locations would be evacuated by chopper and one by the Mobility Troop using Land Rovers.

Cameras and expensive hi-fi equipment was sold to the highest bidder

There was to be no move before 10:00 hours the next morning, so once the tasks had been assigned and kit made ready, most of the men went to the local NAAFI. As everyone tucked into free drink, the NAAFI manager auctioned off all the saleable stock; it was either that or leave it behind. Cameras and expensive hi-fi equipment was being sold to the highest bidder, in most cases for nothing more than a few pounds. There was also a farewell party, at which most people dressed up in white sheets pretending to be Arabs.

Next morning the camp was a hive of activity: several aircraft had already arrived and some RAF personnel were already being loaded. On the main parade ground, several bins containing military documents were being destroyed with magnesium bombs. By 10:00 hours the Wessex helicopter had arrived and an eight-man SAS team took off in it.

A soldier on this extraction takes up the story: 'The two men we were tasked with extracting worked for the Foreign Office (FO). They were stationed in a town 50km (31 miles) to the north of the base. Upon arrival the town seemed fairly quiet, but in this remote region a helicopter was a bit of a novelty. We were dropped off on the roof and, leaving two guys to keep an eye on the local crowd, made our

way down into the courtyard. The two FO guys must have been there for a long time, and by the amount of gin stored in the cellar had intended to stay there a lot longer. Although they had received a message telling them about the withdrawal, they had done no packing or preparation. They fussed around like two old queens, though getting nowhere fast. After an hour the two SAS guys on the roof reported that the locals were getting restless, and that a crowd was gathering. In the end we just dragged the FO guys to the roof. They weren't worried about leaving the documents, but they were depressed about leaving their gin!

10

Left: When VIPs need rescuing from the clutches of terrorists, professional hostage-rescue teams have to be used.

'Eventually all the VIPs were rounded up and taken safely to the base, from where they were airlifted to Aden. By 15:00 hours that afternoon, less than 24 hours after the evacuation order had been received, the camp lay empty, apart from the SAS soldiers. As we had to wait about two hours until an aircraft came back to pick us up, I took the opportunity to explore the deserted camp. The accommodation barracks had been left neat and tidy, with rows of metal beds and blankets stacked to regulation requirements. The small hospital was the same, the surgery well stocked with pharmaceutical drugs and equipment. The mess hall tables stood in neat lines, the chairs stacked on top. The kitchen floor was still wet where someone had mopped it. In the NAAFI yard, pallets of beer sat in the afternoon sun, while the shop looked as if a plague of locusts had been through it. The saddest thing about the whole camp was the Fire Station. Three red tenders stood side-by-side, and one of them was brand new.

During this time much emphasis was put on training and brushing up skills

'At last our aircraft arrived and we piled in, leaving the Land Rovers behind. As we rolled down the runway, we passed thousands of 50-gallon fuel drums, all full. We had wanted to destroy them, but the base commander decided against it. Then, as the aircraft made one last circle, we could see several hundred local Arabs racing for the rich pickings the British had abandoned. I often think about the fire tenders and wonder what happened to them.'

Thus the British presence in Aden ended, and for the next few years the SAS had no real active role. During this time much emphasis was put on training and brushing up individual skills. New departments started to spring up within the Regiment, one such cell being Counter Revolutionary Warfare (CRW). At first it was little more than one officer and a couple of guys sitting in an office at the back of the cookhouse. Their brief was to study the growing terrorist threat and come up with ways of combating the problem. Although the Oman War (1970-76) later came to consume much of the Regiment's time, CRW continued to grow and soon became the hub of all anti-terrorist training, techniques and methods. All types of surveillance were practised, and the SAS learnt good camera skills (including the means of developing and printing the film), as well as infrared photography for night work. Bodyguard courses were organised, which included VIP protection drills and guys being sent on police high-speed driving courses. It was a time of change, and the number of SAS men turning up for work in plain clothes increased.

Four SAS soldiers trained by the CRW wing were dispatched to protect Qaboos

One of the first VIPs the Regiment protected was Qaboos, the Sultan of Oman. His father, the old Sultan Said bin Taimur, had hung onto power as his feudal country fell to rebellious Dhofaris (see Rescuing Our Allies, pp30-53). He refused recommendations from his British advisers to allow the SAS to train loyal Dhofaris in order to fight the rebels. The situation in Oman deteriorated until July 1970, when Qaboos deposed his father. Qaboos had been educated in England and trained at Sandhurst, but on his return to Oman had been kept a virtual prisoner by his father. During the coup, the old Sultan shot himself in the foot and was hastily bundled into an aircraft for a flight to England. Qaboos took control, but the situation was far from stable as some of the older palace guards were still loyal to the old man. Four SAS soldiers trained by the CRW wing were dispatched to protect Qaboos. They remained in close protection for some months, until at last Qaboos could arrange a new trustworthy guard from his own countrymen. One of the original four quit the SAS and remained in Oman to work for the Sultan – he is still there.

For the SAS, the next major threat against VIPs came when a bomb scare occurred aboard the liner *Queen Elizabeth II*. The government

Above: An SAS soldier in Aden, 1966. The Regiment's bodyguarding skills were refined during that conflict.

ordered a small bomb-disposal team to be sent to investigate. Staff-Sergeant Cliff Oliver was senior instructor on the SAS Demolition Wing. The SAS demolition course is the best in the world, but as a consequence of its very nature much of it is still shrouded in secrecy, and remains strictly for SAS eyes only. When the call came, Oliver was instructing students at the demolition bunker, about 16km (10 miles) from Hereford. A chopper whisked him away, and 24 hours later he and three others parachuted from a Lockheed C-130 Hercules into the rough seas. Landing close to the liner, the men were quickly picked up. Although they covered the liner from stem to stern, their search revealed no bomb. Oliver then spent several days enjoying the onboard luxury, and the praise of the rich and famous.

Later that year, the world of international terrorism erupted with a spate of hijackings, murders and kidnappings. In September 1972, the perfect terrorist opportunity arrived as the Germans staged the Olympic Games in Munich. As a result of their treatment of the Jews in

World War II and the nature of the Nazi-controlled Olympic Games held in 1936, the Germans did everything to accommodate the Israeli athletic team, but they also ignored the request of Yasser Arafat, the Palestinian leader, that a Palestinian team should be allowed to compete. A terrorist attack therefore became inevitable, especially as the games coincided with the second anniversary of 'Black September' (when the Palestine Liberation Organisation was expelled from Jordan). The infamous Ali Hassan was tasked with the attack.

The Rome office of Alia, the Royal Jordanian airline, was bombed

At the time Yasser Arafat was keen to build up his reputation as a man of peace – he left revenge to others. One such individual was Ali Hassan Salameh. After the death of Ali Hassan's father, Yasser Arafat cared for the young man, secretly adopting him as his son. Like his father, the young man was a born leader. Together with others of the Al Fatah faction (the Syrian wing of the PLO), Ali Hassan created a new terrorist group known as Black September. It was a secret branch of the PLO with the object of acting independently to take revenge against Jordan. On 28 November 1971, Prime Minister Wasfi Tell, a man very close to King Hussein and opposed to Fatah, was assassinated. He was gunned down in the entrance to the Cairo Sheraton hotel while attending the Arab Defence League Council. As he lay dying, another terrorist lapped at his blood.

Three weeks later, the Jordanian ambassador in London was also shot, though fortunately survived the incident. Then the Rome office of Alia, the Royal Jordanian Airline was bombed, and similar attacks followed in Hamburg, Rotterdam and Bonn. Eventually an airliner was blown-up in midair – Black September was dominating world news. And all the time the thin signature of Ali Hassan signed his mastery over the atrocities, giving him the nickname 'Red Prince'. Pleased with its successes against Jordan, Black September quickly turned its guns against Israel. Taking a leaf from a brother group, George Habash's Popular Front for the Liberation of Palestine, in May 1972 a Black September unit hijacked a Sabena airliner and ordered its captain to land his aircraft at Lod. But the plan misfired: the Israelis stormed the aircraft, killing two terrorists and capturing two more. Ali Hassan went back to the drawing board, and waited for a new opportunity to kill Israelis.

Munich's Olympic village had been deliberately planned with the minimum of security, reducing the unpleasant memories of the past. The Israeli team was billeted in Connollystrasse 31, separated from the public only by a wire fence and the odd patrolling guard. A little before 04:30 hours on the morning of 5 September 1972, a group of young men was seen climbing the fence. This was not an uncommon sight, for many of the athletes stayed out late in the beer halls of Munich. But these were Black September terrorists not athletes, and they were there to seize Israelis and barter their lives for the release of jailed Palestinians.

The Israeli government declined to cave in and refused to free any prisoners

Bursting into the Israeli accommodation, the seven terrorists opened fire, killing many and capturing nine as hostages. The Israeli government, true to form, declined to cave in and refused to free any prisoners. West Germany refused the offer of Israeli troops to counter the problem, and decided to attempt a rescue plan of its own. By 22:00 hours that same night, a grey army bus transferred both terrorists and hostages to the edge of the Olympic village, where they climbed into two helicopters. Next they flew to Fürstenfeldbruck, some 25km (15 miles) to the west, where they landed in a well illuminated area. As requested, a Lufthansa Boeing 727 sat on the tarmac 100m (330ft) away, and two of the terrorists left the helicopter to check it out. At 22:44 hours the first sniper bullet was fired and a fierce firefight ensued. The bolt-action rifles of the German police were no match for the terrorists' AK-47s, and a police sergeant who stepped out from the tower was killed. More police arrived in armoured vehicles, and eventually several terrorists were killed.

Meanwhile, Israeli diplomats watched helplessly from the control tower of Fürstenfeldbruck air base as the inexperienced and ill-equipped German sharpshooters failed to kill all the terrorists in the first volley. Three were still alive, and they fired their guns and detonated their grenades to slaughter the handcuffed hostages as they sat in the helicopters. In the end, the three terrorists were overpowered and captured, but not before they had killed all nine hostages. Waves of shock reverberated around the West, with the massacre seen as both a human tragedy and a warning that terrorism was getting out of hand.

Word started going around that a new anti-terrorist unit was to be formed

Above: An SAS soldier learning some of the many protection drills in one of the specially prepared ranges at Hereford.

(Author's Note: In 1995 I had reason to visit Norway while investigating a terrorist incident. The Israelis had tracked down Ali Hassan to the town of Lillehammer in central Norway. The hit team sent in to kill him identified the wrong man and thus assassinated an innocent Moroccan named Ahmed Bouchiki, who was shot in full view of his pregnant Norwegian wife. I spent several hours with the lady, who has now reverted to her maiden name Torill Larsen. Her story was incredible, and how one of the most sophisticated security agencies in the world could have mistaken her husband for a major international playboy and terrorist is a mystery. Meanwhile, the real Ali Hassan was under heavy security in Beirut, where he had just married the reigning Miss World. On the afternoon of 22 January Ali Hassan left his new wife, at that time six months pregnant, and set off in his distinctive Chevrolet car, sitting in the back squashed between two bodyguards. Several more 'heavies' accompanied him in a white Land Rover. The Chevrolet moved slowly down Verdun Street and into Madame Curie Street. Someone had badly parked a white Volkswagen by the curb, making the street much narrower, and as the Chevrolet negotiated its way through the constriction, the explosives-packed Volkswagen blew up. So, finally, Mossad, the Israeli secret service, had got their man. Ali Hassan was blown to bits, together with one of his bodyguards. Unfortunately several innocent tourists, including a British nurse, also died in the explosion. The woman who triggered the bomb was British. Yasser Arafat helped carry Hassan's coffin, and later comforted Salameh's elder son by the graveside. Months later, on 15 May 1979, his beauty queen wife gave birth to a boy. She called the child Ali Hassan Salameh.)

For the SAS the warning signs were clear, but in the UK police primacy in civil security is absolute, even in Northern Ireland. The massacre in Munich, which had shown the inadequacy of police forces to handle such situations, prompted the prime minister to ask the Director of Military Operations what could be done. The SAS responded with a paper that had been prepared on the subject by an SAS officer called Andy Massey (he had prepared the paper some six months earlier at the request of the SAS's commanding officer). Word started going around the SAS base in Hereford that a new anti-terrorist unit was to be formed. Within days, six SAS soldiers were sent to the Rover factory. Under direct government orders they took the next six white Range Rovers that came off the assembly line. At first the team's duties were fairly basic, but as the enormity of the problem developed and hijackings increased, so did the professionalism of the training. The team was divided into two main sections – the assault groups and the sniper groups – together with a small command and communications group (this structure is still in existence today). Manpower levels depend on tasks and/or the terrorist situation, but in general the minimum strength requirement is about 50 men.

Assault teams focus mainly on forcing entries, concentrating on all the methods of getting into the operational area, be it an aircraft, train or building. Snipers, on the other hand, deal with any long-range situation that may present itself. The two teams exercise independently, but a lot of cross-training also goes on, thus providing the numbers to suit the situation required.

The 'Killing House' was built with the purpose of perfecting shooting skills

All members of the SAS anti-terrorist team spend hundreds of hours in the now famous 'Killing House'. This purpose-built, flat-roofed block building, located in the grounds of the Hereford base just behind the old quartermaster's stores, is unique. It was designed and built with the express purpose of perfecting the individual shooting skills of SAS personnel, and allows for many different sets of circumstances. From the moment it was completed the building has been in constant daily use, not just by the anti-terrorist team practising room assaults, but also by the CRW wing to train men in bodyguards' drills. The SAS became good at these tasks through hard work, innovative skills and the development of new procedures. This level of specialist training has its drawbacks, not least the fact that at one stage the Regiment was in constant demand to host VIP visits, all of them time-consuming. The positive side of these visits, though, was the fact that many of the VIPs realised how advanced the SAS had become, and subsequently many of their personal bodyguards were sent to Stirling Lines for training.

Running parallel to CRW and the anti-terrorist team, a small but vitally important cell also operates within the SAS. This is Operations Research (OR) and comprises two men, normally a sergeant and a corporal, tasked with finding or developing the specialised equipment needed by the Regiment. This ranges from weapons, clothing and vehicles to food rations. Any member of the Regiment can approach OR and request information or put forward an idea. OR then searches the planet to accommodate that need if it is deemed necessary. Rations have always been a bone of contention within the SAS. Time and again they have been changed or modified to make them lighter, give more energy, and have more taste and texture. At one meeting several SAS soldiers sat round a table with top food scientists from Bath. 'We want a ration that gives 3500 calories but weighs only 1kg,' demanded the SAS soldiers. 'Fine. We can give you half a kilo of fat and half a kilo of sugar,' came the reply. Eventually this type of give-and-take bartering process resulted in the procurement of the very best equipment, much of which later filtered down to general Army units.

When a request was put to OR for a device that would disorientate a terrorist without harming any hostages in close proximity, OR in turn asked the technical experts at the Royal Small Arms Factory at Enfield to help. Of the several devices that were made, the most popular was the 'stun grenade'. Among the others was the 'screamer', a little beauty that emitted a very loud scream. The problem was that although it would unnerve the terrorists, it was also capable of giving the hostages and the SAS assault team heart attacks!

Many royal VIPs felt they could travel around the UK with little security

OR has come up with some brilliant equipment over the years. Small, one-man aircraft with motorised propellers stuck on the back and supported by a parachute could often be seen flying around Hereford, while on the ranges strange-looking space-age crossbows could be seen penetrating 10mm (0.5in) metal at 100m (330ft). Even run-of-the-mill items, like socks, would be purchased and distributed to the squadrons for trial. In the end, only that equipment which did the job perfectly was recommended for purchase.

In the early 1970s, many royal VIPs felt they could travel around the UK with little or no security. True, many had bodyguards, but at the time they were chosen for their capability to react adequately in social circles rather than to react quickly and efficiently against a terrorist incident. And at times many of the royal family would drive from Buckingham Palace to

Windsor without any escort or back-up. With terrorism running wild in Europe, it was only a matter of time before there was an incident. When it came, the targets – Princess Anne and Captain Mark Phillips – were lucky: their attacker was a straightforward mental case. Had he been a terrorist, the UK would have been minus one princess.

On the night of 20 March 1974, Princess Anne and Captain Phillips were driving down the Mall, London, accompanied by Inspector Jim Beaton. Beaton was of the old school: neat appearance and well educated. On this occasion he was armed, but unfortunately his gun was defective. He had joined the Royal Protection Group a year earlier, and had been fully trained on the 9mm Walther PPK pistol.

(Author's Note: Despite its limitations regarding the number of rounds carried in the magazine, the Walther PPK is a slim-fitting weapon and has often been used by the SAS. I carried one for years in Northern Ireland because it is well suited to operations in civilian clothes: it can be concealed more effectively than the bulky Browning High Power. Used with precision, it is one of the finest bodyguard weapons around. Its main fault, if it has one, is the magazine. If this is not emptied on a regular basis, thus allowing the spring to relax, stoppages will occur.

At the time the Royal Protection Group, which consisted of about 20 men, fired little more than a few hundred rounds during training, and refresher training consisted of

Below: Sultan Qaboos of Oman, who was put under close SAS protection as a consequence of his coup in July 1970.

firing little more than 50 rounds. In comparison with the training of the SAS bodyguard group, this training was minimal. As any good shot will tell you, the more you shoot the more proficient you become.)

On the night in question, the royal party was returning from a charity reception in the city. By 19:59 hours its somewhat old Rolls-Royce limousine was driving up the Mall, just minutes away from Buckingham Palace. At this stage a white Ford Escort deliberately cut in front of the Rolls-Royce, forcing it to stop. There are several different versions about what happened next.

After he had shot Beaton, Ball then tried to open the driver's side rear door

The bodyguard, Beaton, thinking it was nothing more than a typical London lunatic driver, left his front passenger seat and walked around the rear of the car, taking up position by the offside rear door. This put him between the driver of the white Ford and Princess Anne. No sooner had he stopped to survey the situation than a shot rang out, hitting him in the chest. Staggering backwards, Beaton immediately sought shelter behind the Rolls-Royce. Instinctively he drew his pistol and fired. There is some argument as to how he did this, but in any case he missed. The princess is later reported to have said the following: 'The policeman got off one shot, which I am convinced came through the back window of the car as something hit me on the back of the head – so I thought that was a good start.'

Beaton, although seriously wounded and quickly loosing blood, made another attempt, but this time the gun just went 'click'. In normal circumstances, the correct stoppage drill would have cleared this in seconds. Beaton was going through the right motions mechanically, but was clearly not switched on: because he was wounded his ability to operate was greatly reduced. The chauffeur, Alexander Callender, had seen Ian Ball, the driver of the Ford, deliberately smash into the Rolls-Royce. He had also seen Ball step out of the car and produce a gun. After he had shot Beaton, Ball then tried to open the driver's side rear door where the princess was sitting. Both the princess and Captain Phillips held onto the handle and a tug-of-war ensued. Ball finally managed to get hold of Princess Anne's arm and tried to pull her from the vehicle; the princess resisted by hanging on to her husband. When, at last, the sleeve of her dress came away in Ball's hand, they managed to get the door closed once more.

Beaton made another attempt at confronting Ball but, seeing his advantage, the latter demanded that the officer put down his weapon. Beaton did so, but then climbed into the car to sit by the princess, an act of defiance if nothing else. The lady-in-waiting, Rowena Brassey, who had accompanied the royal couple, left the vehicle and went to pick up the officer's weapon. It was a gesture of derring-do, and an honourable one at that, but cries from the crowd which was now gathering persuaded her not to. At that moment a journalist, John McConnell, having heard the shots fired, arrived on the scene and challenged Ball. The latter's response was to produce a second weapon, a 0.22in pistol, and shoot McConnell in the chest. The journalist staggered away, and Ball concentrated once more on his efforts to kidnap the princess. He repeatedly threatened Beaton, pointing the pistol at the window, directly at his head. Instinctively knowing that Ball was about to shoot, Beaton put his palm to the window and Ball shot him in the hand. Had Beaton not acted so, the bullet would have certainly hit either him or the princess.

Beaton lashed out with his foot, and Ball shot him a third time

A policeman, PC Michael Hills, who had been on duty at nearby St James Palace, had also responded to the shootings. He radioed Cannon Row police station, and at last the authorities were alerted. There had been no radio communications system fitted to the Rolls-Royce, and personal radios were thought to be intrusive by the Royal Protection Group at that time. Arriving at the scene, Hills confronted Ball, only to be shot in the stomach. As he crawled away, the chauffeur Callender opened

Above: Staff-Sergeant Cliff Oliver and his team descend by parachute into the ocean prior to their search of the *Queen Elizabeth II*.

his door, and Ball shot him in the chest. By now the situation was becoming critical, and Ball tried once more to open the rear door. As he succeeded, Beaton lashed out with his foot, and Ball shot him for a third time. On seeing this, several have-a-go merchants descended on the gunman. One positioned his car to block Ball's vehicle. Another man, Ronald Russell, who had been passing in a taxi, tackled Ball and managed to punch him. Ball fired at him but missed. Russell was still fighting with Ball when the police arrived. Ball made a run for it but was soon tackled by a policeman and captured.

Ball was locked away for the duration, and a full inquiry was held into the incident. One cannot pin blame on men who have been wounded in the line of duty, and just rewards were rightly meted out to those involved. The 9mm Walther seemed to be the culprit, and was subsequently replaced by the Smith & Wesson 0.38in Special. Other improvements were made, such as the installation of radios in royal vehicles, which were also fitted with bulletproof glass and automatic locks. The most important change came when the head of the Royal Protection group, Commander Trestrail, visited Hereford in early 1975. He was not given the normal demonstration, but was instead party to a three-day VIP exercise. It soon became abundantly clear to Trestrail that very high standards of training already existed at Hereford which could benefit royal VIP protection. After this the Royal Protection Group all came to Hereford for training.

It is not just the British Royal Family which has taken advantage of the SAS, for the VIP protection groups of many friendly nations have, over the years, also been trained by small specialist groups from Hereford. Their purpose has not been to supply permanent bodyguards, but to train the local forces to do the job. Since 1975, moreover, many ex-SAS soldiers have become bodyguards for the rich and famous, and although the demand has declined it is still the largest single industry for ex-Regiment personnel. The OPEC conference hijack by the terrorist Carlos in 1975 did much to initiate this, and soon afterwards the oil ministers who had been kidnapped hired SAS bodyguards. At one stage it was so fashionable to have an SAS bodyguard that the drain on the Regiment was excessive. For many there were rich pickings and an exotic lifestyle to be had.

Rose learnt that the small nation of the Gambia was in the throes of a coup

The term of Margaret Thatcher (1979-90) saw a huge increase in security, and one of Mrs Thatcher's first changes was to order several armoured Daimlers. She also increased the number of policeman for the protection of top politicians. The SAS was directly tasked with providing training for this increased manpower, and for a while Hereford became the 'Bodyguard Mecca'. In special circumstances, where the risk was thought to be abnormally high, the SAS took over the protection task itself. Soviet defectors and A-listed Ministry of Defence were VIPs requiring SAS bodyguards.

And then there were the 'special' jobs, those little tasks that drop out of nowhere and fit so neatly into the SAS style. One such incident was the rescue in Gambia.

On 1 August 1981, Lieutenant-Colonel Michael Rose, then commanding officer of the SAS, was walking in the Welsh countryside. It had been a busy time for the Regiment, which had recently been drafted in to help with the protection of numerous VIPs. A large number of Commonwealth heads of state had been in London, attending the wedding of Prince Charles to Lady Diana Spencer. The gathering of notables had presented a number of opportunities for terrorists, but now that the wedding was over the SAS men could return to Hereford and relax a little. Rose therefore arranged to take time off to enjoy the company of his children, something rare for the SAS commander. Even now he wore an electronic beeper on his belt, and this interrupted Rose's much-anticipated interlude with an urgent summons. With all haste he made his way to the nearest telephone and made a call to Hereford headquarters. Diane, a telephonist of long standing in Hereford, switched his call to Major Ian Crooke, who was acting executive officer at the time. From Crooke Rose learnt that the small nation of the Gambia, a former British colony, was in the throes of a coup d'etat. Launched two days earlier, it coincided with the absence of the nation's president, Sir Dawda Jawara, at the royal wedding. Crooke reported that Jawara had asked Mrs Thatcher for help. Senegal, the Gambia's neighbour, had already sent troops to combat the rebels, but Thatcher agreed to dispatch a couple of SAS men to the scene under the strictest secrecy. Even this modest response, if it became public, could lay her government open to charges of renewed imperialism in Africa.

Rose told Crooke to pick a man to go with him and get to the Gambia

Rose was admired throughout the UK as the commando leader who had freed 20 hostages in the brilliant 11-minute rescue operation at the Iranian Embassy in London during the preceding spring. Speaking on the telephone with Crooke, Rose was inclined to take the assignment himself. Not only was he the boss, but he had a reputation for being almost prescient at making quick decisions without benefit of all the facts, a skill that might well prove to be useful in the Gambia. Moreover, the prime minister herself had called on the Regiment: its reputation was at stake. But there were problems with the commander himself going: he would have to be picked up by helicopter, and someone would have to come to get his children and his car. All of that would take several hours, and in these situations even minutes can be crucial. So Rose told Crooke to pick a man to go with him and get to the Gambia on the first available flight.

The Gambia is almost completely surrounded by Senegal, a former French colony. It is among the smallest nations in Africa, and one of the continent's few genuine democracies. In 1980, its average annual income – only $210 per capita – was on the decline because the local cash crop, peanuts, had fared poorly in two years of unusually dry weather. Tourism, an important source of revenue, was also on the wane. Much grumbling could be heard among the populace over escalating prices for rice, cooking oil and sugar, and also about the high rate of unemployment. The disaffected went about painting anti-government slogans on walls. The president's private yacht mysteriously caught fire, and for several months his advisers had told him of their suspicions that a coup was being planned. Jawara, however, discounted the information and flew to London for the royal matrimonial festivities.

Muscle for the attempt to overthrow Jawara was provided by Usman Bojang

Trouble in the Gambia at this time automatically raised the spectre of Muammar Gaddafi. The mercurial Libyan leader envisaged a confederation of Islamic African states under his guidance. To this end he attracted exiled African political leaders of Marxist persuasion to the Libyan capital, and plotted with them the reshaping of a number of African governments. With Libya's oil wealth, he purchased Soviet-bloc weaponry to arm his troops. He dispatched forces to shore up the brutal Ugandan president, Idi Amin; he ordered his army into Chad; and he supported Polisario guerrillas fighting Morocco for control of the Western Sahara. The Gambia, being 70 per cent Moslem, lay within Gaddafi's fancied Islamic realm.

At 05:00 hours on Thursday 30 July, the coup erupted. Muscle for the attempt to overthrow Jawara was provided by Usman Bojang. A former deputy commander of the Gambia's 300-man Police Field Force, a paramilitary organisation charged with preserving order in the tiny country, Bojang managed to persuade or force the contingent based in the town of Bakau to join the coup. This group, which amounted to about one-third of the organisation, disarmed most of the loyal police, then quickly took over the nearby transmitter for Radio Gambia and then moved into Banjul, the capital. On the way, the group opened the country's largest prison and distributed weapons from the police armoury not only to the inmates but to virtually anyone in the immediate area. Not long after daybreak, citizens and former prisoners alike began rampaging through the streets and looting shops. Soon, a free-for-all erupted. Within the first few hours of the coup, scores of bodies – policemen, criminals and civilians – littered the streets of Banjul.

Ideological backing for the revolt was provided by the Gambian Socialist Revolutionary Party, which was headed by a young Marxist named Kukoi Samba Sanyang. His

Below: Sir Dawda Jawara, President of the Gambia, whose rule was threatened by a coup that broke out on 30 July 1981.

Above: The ferry that was used by Clive Lee, an ex-SAS veteran, to reach Banjul during the Gambian crisis, as described in this chapter.

given name was Dominique, but when he became a communist he changed it to Kukoi, a word in the Mandinka language, native to the Gambia, that means 'sweep clean'. Sanyang was also one of the African radicals who had spent time in Libya.

Arriving at Radio Gambia shortly after rebel policemen had seized the station, Sanyang closed the country's borders and its airport at Yundum, some 24km (15 miles) east of Banjul. He then proclaimed a 'dictatorship of the proletariat' and charged the 'bourgeois' President Jawara government with corruption, injustice and nepotism.

Very few foreigners lived in the capital. The majority resided in the nearby communities of Bakau and Fajara, where most of the tourist hotels were situated. Europeans and Americans – some of them on vacation, others working in business or government capacities – generally stayed off the streets. In some cases, rebel Gambian policemen, who saw no profit in harming such individuals, guided anxious foreigners to the residence of the United States ambassador, Larry Piper. The house was soon haven to 123 nervous guests, 80 of them American citizens.

On the outskirts of Banjul, along the nation's beautiful coast, a number of European tourists had sought shelter in the Atlantic Beach Hotel. Two armed looters found their way inside, ransacked the safe, took the hotel

manager hostage and fled the building – right into the gunsights of unidentified adversaries, who shot and killed the two criminals virtually on the hotel doorstep. Hearing more firing outside, the foreigners shrank away from windows. They organised watches and even posted guards, 'armed' with fire extinguishers, at the hotel's entrances.

Had such stories reached the world outside the Gambia, the coup would have attracted far more attention than it did. During the first day of the coup, however, Jawara was in effect controlling the news from London. In contact via telephone with his vice president, who had taken refuge in the Banjul police headquarters under the protection of loyal troops, Jawara wisely made himself accessible to the press. By doing so he was able to play down events at home. As if to resume control of the government, Jawara boarded a jet bound for Dakar, capital of Senegal. Before leaving, he acknowledged that he had considered invoking a mutual assistance treaty that Gambia had signed with Senegal some 15 years earlier to fend off external aggression. That no foreign provocateur had yet surfaced seemed to be of little concern to him.

Such was the situation as Rose ordered Crooke to the Gambia. The major and the SAS sergeant he chose to accompany him assembled their gear: 9mm Heckler & Koch submachine guns, 9mm Browning High Power automatic pistols and a large stock of ammunition and grenades. The first available transport to the area happened to be an Air France commercial flight to Dakar.

Crooke managed to pass his little arsenal through customs and baggage checks

Although firearms can be carried in checked baggage aboard such aircraft, explosives are prohibited. By availing himself of some informal contacts within British and French military and diplomatic circles, however, Crooke managed to pass his little arsenal through customs and baggage checks and onto the plane. Dressed casually in blue jeans, the two SAS men drew no attention from their fellow travellers, among whom were many reporters and television camera crews.

The day after arriving in Senegal, Crooke encountered his first obstacle in the form of British diplomats. With fighting going on between the rebel forces and Senegalese troops, and with a considerable number of British citizens in danger, officials in both Dakar and Banjul decided that it would only complicate their duties further if Crooke and his sergeant were to get into the act. Preferring simply to applaud the Senegalese if the intervention succeeded or to chastise them if it failed, in true Foreign Office style the diplomats forbade the major and his sergeant to proceed.

Crooke listened impassively to these instructions, then returned to the airport and began looking for the quickest way into the

Gambia. The major reasoned that if the prime minister had sent word to detain him, the harried British diplomats in Senegal would have said so. And with no word from Mrs Thatcher, at whose behest he had come to Africa, Crooke decided to go ahead with his mission despite the remonstrations of these relatively minor officials. He resolved to hop on an aircraft bound for the Gambia's Yundum airport, where the Senegalese paratroop commander, Lieutenant-Colonel Abdourah-man N'Gom, had established his headquarters.

Crooke set about translating resolve into action. While he was doing so, an ex-member of the Regiment was embarking upon his own adventure. Clive Lee, a very tall and hulking retired SAS major, was employed in the Gambia as a civilian adviser to the Gambian Pioneer Corps, a division of the Field Force that trained rural youth in agricultural and construction skills. Hearing of the coup on the radio, Lee had rounded up 23 Pioneer Corps members, armed them and set out for Banjul. To get there from the Pioneer Corps base in the town of Farafenni, which was 100km (62 miles) east of the capital, he had to cross the River Gambia. Because of the hostilities, however, the ferry was not running. The sight of the Briton and his men persuaded the captain to rouse his crew and take Lee's band to the other side.

Senegalese paratroopers entered Banjul after a fierce battle

Once across the river, they made their way to Banjul, moving through mangrove swamps to avoid rebel positions along the main approach to the city. Reaching the capital on 1 August, they headed for the police headquarters to reinforce the small unit of loyal troops, and barricaded nearby streets to deter the rebels.

That evening Senegalese soldiers entered Banjul, having captured Yundum airport a day earlier after a fierce battle in which nearly half of the 120 paratroopers making the assault had been wounded or killed. Within a few hours the Senegalese had cleared Banjul of rebels, and had gained control of Denton bridge across Oyster Creek to prevent insurgents from re-entering the capital from their concentrations in Bakau and Fajara. With the route from Banjul to the airport reasonably secure, Lee had set off for the airport and Dakar.

The SAS keeps in contact with former members, so Crooke probably expected to come upon Lee at some point. It is even conceivable that Lee had word that Crooke was en route and hurried to Dakar to meet him. In any event, the Pioneer Corps adviser was quickly recruited. Introducing themselves to one of Lieutenant-Colonel N'Gom's officers, the three Britons had little trouble finding a space on an aeroplane bound for the Gambia.

Three Britons carrying submachine guns could hardly escape notice

On arrival, they found the situation little changed. Although N'Gom continued to strengthen his forces in Gambia and occasionally traded shots with the rebels, the military situation had reached an impasse. His troops were stalled outside Bakau because Sanyang had taken more than 100 hostages. The most valuable captives were Lady Chilel N'Jie – one of President Jawara's two wives – and a number of his children. In addition, Sanyang held several members of the Gambian cabinet. And although N'Gom had wrested Radio Gambia from Sanyang's rebels (the transmitter lay between the airport and the bridge), the coup's leader had commandeered a mobile transmitter, from which Lady Chilel appealed almost hysterically to Senegal, announcing that the hostages would be executed unless the paratroopers withdrew. Sanyang repeated the threat himself: 'I shall kill the whole lot,' he warned, 'and thereafter stand to fight the Senegalese.'

On 5 August, Crooke decided to make a reconnaissance. The blue-jeans-clad SAS officer and his two associates slipped out beyond the Senegalese outposts and set out on foot. The weather was hot: during August in the Gambia temperatures routinely exceed 30 degrees C.

Three Britons carrying submachine guns could hardly escape notice in Fajara, but the outing was not as dangerous as it might seem.

Although there was always the chance that an encounter with an armed insurgent could end in gunfire, the rebels did not seem inclined to harm Europeans. Furthermore, Crooke observed an unmilitary laxity among the troops manning the rebel positions. Unknown to both him and the outside world, Bojang had been killed during the second day of the coup. His absence, and the resulting lack of leadership, probably accounted for the apparent decline in rebel vigilance. Crooke's sortie confirmed that the lightly armed insurgents were now capable of little more than token resistance against the well trained Senegalese troops.

Crook persuaded N'Gom to begin an advance to Fajara and Bakau the same day. The British officer and his companions accompanied a contingent of Senegalese troops along the hot byways of the suburbs. Peter Felon, a British engineer employed by an American crane company, saw the party when they appeared at his hotel in Fajara. 'Ten Senegalese troops and a British army officer arrived at the hotel,' the engineer recalled. The officer, probably Clive Lee, wore khakis with no insignia. 'With him were two men who I can only describe as the most vicious-looking professionals I have ever seen.' Upon being told that rebels were hiding along a creek near the beach, the pair set off to find them. 'There was sporadic violent gunfire,' said Felon, 'then the two men walked calmly back to our hotel.'

An American aid worker at the US Embassy was one of the foreigners who had taken refuge in Ambassador Larry Piper's house. It was early afternoon, he remembered, when a lookout they had posted announced that soldiers were coming up the hill. 'The house,' the aid worker later said, 'was on a bluff sloping to the beach. I went out and saw a wave of Senegalese come running up the hill in full camouflage-type gear led by three whites, one of whom had on an Australian hat, khaki shorts, and a knife strapped to his leg. It was literally like living in a movie.' After ascertaining that everyone was all right and leaving behind a dozen or so paratroops for security, the party disappeared.

Arriving at the British high commissioner's offices, Crooke learned that armed rebel guards had escorted President Jawara's wife and her

Below: The antenna of Radio Gambia, the station which was seized by the rebels on the first day of the coup.

four ailing children – one of them an infant of only five weeks – to a British clinic the day before. Doctors at this tropical disease research facility, which stood only a block or two from the high commissioner's office, treated the children and advised her to bring them back within 24 hours. The interval had passed, and now Lady Chilel had returned for follow-up care. This information came by way of a telephone call to the high commissioner from the British physician attending the children. The official told the doctor that armed SAS men would be there within minutes. Meanwhile, Crooke and his two companions quickly headed for the hospital.

Hearing that help was on the way, the doctor began to draw out his treatment. The wily physician even convinced the woman's armed escorts that they were frightening his other patients, and persuaded them to put their guns out of sight.

The two guards froze when they felt gun muzzles at the back of their heads

As Crooke approached the hospital, he noticed two armed guards posted at the entrance. Handing his submachine gun to his companions, the major instructed them to circle behind the guards. He then walked up to the two guards and distracted them in conversation as the other two SAS men crept up from the rear. It is difficult to imagine what Crooke could have said or how devious a plan he might have formulated in order to draw the guard's attention away from his accomplices. The SAS is mute on the topic, following the lead of the British and Gambian governments in their steadfast refusal to acknowledge that the UK had sent any military assistance at all. But whatever Crooke's ruse, it worked. The two guards froze when they felt gun muzzles at the back of their heads.

Leaving the captives in the hands of his able assistants, Crooke slipped inside the clinic. He surprised Lady Chilel's weaponless escorts as they watched the children being treated, and promptly took them prisoner. After conducting the president's wife and children to the high commissioner's office, Crooke and his party retreated to Lieutenant-Colonel N'Gom's headquarters at the airport.

A day earlier, Senegalese troops, who now numbered about 1500, had found and destroyed the mobile transmitter that Sanyang had been using. Although Sanyang himself escaped, he was no longer a factor in the situation. With the silencing of its leaders, the coup's backbone was broken. Yet many hostages remained under rebel guns at a police barracks, and bands of turncoats, disaffected policemen and criminals had yet to be rounded up.

The only confirmation of the SAS presence came from a Senegalese officer

N'Gom paced his advance slowly. Panic among the rebels might cause them to begin killing their prisoners. They had nearly done so a few days earlier, when a policeman who had been forced against his better judgement to join the coup began shooting some of the rebel guards. He was killed at once, but his brave act seemed to thwart the planned execution. The Senegalese paratroops edged up to one side of the barracks, leaving several exits unguarded for the rebels to escape from. After a tense hour the hostages walked free and the insurgents dispersed, only to be captured later.

It was all over. After eight days of rebellion and perhaps as many as 1000 deaths, President Jawara was once again the unchallenged and elected head of the Gambian government. In Banjul he posed for reporters, hugged his baby son and pronounced: 'I'm relieved and happy.' However, when asked about the Europeans his wife Chilel had said had rescued her and her children, he stonewalled, implying that she and any others making such an assertion must have been mistaken.

Crooke stayed on in the Gambia only long enough to satisfy himself that the British citizens were safe. A number of whites besides Peter Felon had observed him and his two companions carrying submachine guns, but the SAS men refused to reveal their identity or acknowledge their connection with the British Army. To a group of happy Canadians they

Above: The building used by Major Ian Crooke to gain access to the hospital where Lady Chilel was being held hostage.

admitted being British, but not SAS. 'We don't say anything about that,' was their reply.

To make Jawara's political recovery as easy as possible, all official comment about the counter-coup and rescue was reserved for African governments. The only confirmation of the SAS presence came from a Senegalese officer, who told reporters that SAS personnel had indeed participated in restoring order. True to form, the British government has remained silent to this day.

Kukoi Sanyang was eventually arrested in the neighbouring country of Guinea-Bissau, but the socialist government there later released him, despite Gambian requests for his extradition. Senegalese troops captured more than 100 of the rebels and convicts, seven of whom were ultimately condemned to death. Libya was never directly connected with the coup attempt.

The SAS operation in the Gambia was a gem of its genre: situations in which a few skilled and confident soldiers can influence an event far out of proportion to their numbers – and depart leaving almost no trace of their presence. Crooke was so discreet that his name was not linked with the episode for almost seven years. The British government, by refusing to admit any involvement, handled the affair as an African problem which was solved by Africans. Partly as a result of London's smooth state-craft, neither Libya nor the USSR could make political gains in West Africa during the early 1980s.

The success of the action, however, hinged in large measure on Crooke's initiative and good judgement. That he had a lot of help is undisputed. It came partly in the form of militarily incompetent coup engineers and partly from the presence of the Senegalese. N'Gom's troops supplied the manpower needed to fight rebels who were disinclined to surrender, and also to maintain security in areas that had been swept clean. Nevertheless, Crooke was the one who tipped the scales. He chose to ignore the restrictions placed on him by British diplomats and acted as he believed the prime minister wished. Had he been wrong, he would have no doubt suffered severe consequences. An SAS officer is expected to act boldly, and Crooke could not face his commander with the excuse that timid envoys had prevented an obvious course of action.

The Thatcher years saw a growth in the bodyguard industry and in the British quick-reaction capability to deal with terrorist events. Most of the manpower was drawn from the police, but in some areas civilian firms provided the protection. These firms, almost entirely made up of ex-SAS men, carried out the overseas tasks such as the protection of embassies and British diplomats in high-risk areas like the Middle East. But terrorists do not give up easily, and despite the millions of pounds of public funds spent on security, Prime Minister Thatcher was given a reminder of the seriousness of the threat to VIPs.

In October 1984, the Irish Republican Army (IRA) made a serious attempt to wipe out the entire British cabinet at the Tory party conference in Brighton. It was the last day of the Conservative Party's four-day annual conference. The prime minister, 13 of her 20 Cabinet colleagues and many of her senior advisers were staying at Brighton's venerable Grand Hotel, a Regency-style building on the sea front. She had just put the finishing touches to her speech when, at 02:45 hours, the night's silence was shattered by a thundering explosion. A powerful bomb had detonated four floors above the cluster of suites occupied by the prime minister and some of her colleagues, blowing out a section of the nine-storey building's facade. The resultant hole was 9m (30ft) deep and 4.5m (15ft) wide, and the explosion sprayed broken glass and chunks of concrete through the halls and onto the street.

Tons of plaster, flooring and furniture crashed from floor to floor

The gap extended from the roof to the fifth storey. Tons of plaster, flooring and furniture crashed from floor to floor, finally destroying the Grand's elegant foyer, where Tory leaders had gathered only hours earlier. Thatcher's suite, located only 9m (30ft) below the source of the blast, was badly damaged, and its bathroom was totally demolished. Miraculously, the prime minister was unhurt. 'This conference will go on as usual,' she declared firmly as she emerged from the wreckage, accompanied by her husband, Denis. Thatcher was fully dressed, her face flushed with anger. 'We were very lucky,' she said, putting on a brave face. Inwardly, though, the Iron Lady was furious, as well as relieved.

Others were not so fortunate: four people were killed and at least 34 others injured. Among the dead were Sir Anthony Berry, a former Tory deputy chief whip; Eric Taylor, chairman of the Northwest Area Conservative Association; and Mrs John Wakeham, wife of the chief whip. Wakeham himself was injured, as were Alfred Parsons, the Australian High Commissioner to the UK, Norman Tebbit, the Trade and Industry Secretary, and Tebbit's wife. Wakeham lay buried for nearly seven hours before being rescued. Tebbit, who subsequently underwent exploratory surgery to determine the extent of his severe chest injuries, spent over four hours under the rubble.

The IRA had in effect tried to destroy the entire British government

Some nine hours after the blast, the Irish Republican Army claimed responsibility. In a telephone call to the Irish state radio in Dublin, the group asserted that it had set off a gelignite bomb in an attempt to kill 'the British Cabinet and the Tory warmongers'. The IRA promised more violence in the future. 'Thatcher will now realise,' it said, 'that Britain cannot occupy our country, torture our prisoners and shoot our people in their own streets and get away with it. Today we were unlucky. But remember, we have only to be lucky once. You will have to be lucky always.' It was the boldest and most outrageous strike ever against public officials in Britain. The IRA had in effect tried to destroy the entire British government.

The scene in the hours immediately following the Brighton blast was one of devastation. As Thatcher and her husband were taken to the safety of a police station, the residents of a nearby hotel, including Charles

Right: The damage caused by the IRA bomb at Brighton in October 1984, which nearly killed Prime Minister Margaret Thatcher.

H. Price II, the US Ambassador to Britain, were evacuated for fear of a second attack. Working with the help of TV lights and from time to time calling for quiet so they could hear cries for help, rescue workers used axes to chop through the debris, and brought in a crane to reach those trapped on high floors. Shocked delegates wandered along Brighton's sea front promenade in their night clothes, while dishevelled cabinet ministers worried about losing government papers. Lord Gowrie, Minister for the Arts, dragged canvas deck chairs from the beach for use as makeshift stretchers. Education Secretary Sir Keith Joseph, in pyjamas and silk robe, sat on his dispatch case by the shore.

Police began an investigation into the bombing and the security failure

Thatcher had had a narrow escape. A Special Branch officer on the security detail in her hotel said she had left her bathroom two minutes before it was destroyed by the blast. Tory Party chairman John Selwyn Gummer also had a close call. 'I was just outside Prime Minister Thatcher's suite when I was thrown backwards by the force of the explosion,' he recalled. 'The Prime Minister came through the door and the first thing she said was "Is there anything I can do to help?" She was totally calm and looked very angry.'

The scale of the attack and the audacity of the IRA far exceeded their previous attacks in mainland Britain (a dual attack in the heart of London on 20 July 1982 killed four troopers of the Household Cavalry in Hyde Park together with seven horses, and seven members of the Royal Green Jackets' Band in Regent's Park; exactly one week before Christmas the next year, a car bomb went off outside Harrods department store in London, killing six people and wounding another 94).

Police immediately began an investigation into the Brighton bombing and the apparent failure of security measures. Experts of the anti-terrorist squad announced that the bomb had probably weighed 9kg (20lb) and not 45kg (100lb) as claimed by the IRA, and that it was triggered by a timing device. The fact that this had been programmed weeks in advance indicates the expertise of the bomb maker. The technology is simple and available in most households: how often do we pre-programme a video recording of a TV show we wish to watch in the future? It is just a matter of connecting a microswitch to the device and, instead of recording the late night film, a bomb is detonated instead.

As the inquiry got under way, immediate questions arose as to how tight security had been at the hotel. Journalists reported they were able to enter without having their briefcases

examined. Some government officials believe the bombers walked into the hotel openly with stolen passes, blending in with delegates and party officials. At the week's end, however, a Scotland Yard expert disclosed that the bomb was a highly sophisticated device that could have been planted weeks ago in the floorboards of the sixth-floor room.

It turned out that the only security check for explosives was carried out by a police sniffer dog. Despite the animal's complex sense of smell, the process of wrapping the explosives in clingfilm and leaving the package outside for a week will negate any odour.

Above: Yet another ghastly IRA atrocity. The bomb planted in Hyde Park killed a total of four soldiers and seven horses.

The Brighton bombing was a timely reminder of the need for close protection of VIPs, and for a lot of forethought when it comes to screening venues ahead of their use by politicians and royalty. The Regiment played its own part in the aftermath of the bombing, stepping up its campaign against the IRA. Thanks to SAS training, the Royal Protection Group and other police units around the country are better equipped to safeguard VIPs.

WHEN RESCUES GO WRONG

Even the best planned actions can go wrong, but when it does, elite teams such as the SAS are taught not to panic but to keep thinking, and retrieve the situation as best they can.

Aden became so unstable in the mid-1960s that the British finally decided to pull out altogether. Once this decision had been announced, things deteriorated and the rebels started attacking any British people they could find. One December morning in 1967, the town of Crater (so named because it sat in the base of an extinct volcano) went ballistic. The rebellious locals killed almost every white man, woman and child that could be found, and rumours of terrible atrocities quickly spread.

An SAS soldier serving in Aden at the time takes up the story: 'I was sitting in Ballycastle House, near Khormaksar airport, which served as the SAS operational headquarters. I had been lying on my bed when the Boss stuck his head around the door and yelled, "Move. Move now." This may have been my first year in the SAS, but I was rolling. I grabbed at my belt kit, already complete with ammunition and water. Then, shouldering my rucksack and grabbing my L42 sniper rifle, I ran for the chopper which was waiting

Left: Operation 'Eagle Claw' gets underway. Several Sea Stallion helicopters take off from the carrier USS *Nimitz* on their way to Desert One.

Above: Former Italian Prime Minister Aldo Morro, who was murdered by Red Brigade terrorists in 1978.

outside on the pad. Here a group of us had assembled, and we stood in anticipation while the boss made his selection. Looking at the L42 sniper rifle in my hand, he pointed at me: "You, on the chopper. All Crater is hostile."

'Minutes later we were on the Jebel Shamsan, the high mountain that looked down over Crater. Once the noise of the chopper fade away, Jock, the senior officer, gave us the big picture. It would seem that the rebels really were slaughtering all the whites, and our task was to shoot anyone who was carrying a weapon. That included the local town police, who had sided with the rebels. Our patrol consisted of six members, all troopers, and made its way into a good defensive position overlooking the town. Three of us were armed with sniper rifles, which, as things turned out, was the ideal weapon for the situation, as most targets ranged from 300-600m (990-1965ft).

'Over the years I have become used to death, but in those days I was a novice and totally unprepared for what I saw. The bodies of butchered men, women and children had been laid out neatly in the street, positioned to allow traffic to run over them. Yes, I kept my cool – just. We settled down and started selecting targets, after about 10 minutes nothing moved on the streets. Next day, Lieutenant-Colonel "Mad Mitch" Mitchell, against the advice of the High Commissioner, bravely led the boys of the Argyll and Sutherland Highlanders into Crater and retook the town. Rumour has it that the local police had cleaned up the bodies and placed them all in tea chests before handing them back to the British, but this was never confirmed.'

Despite a monstrous beating by the IRA, Nairac did not talk

It has not always been possible to rescue men in danger. One such instance was that of Captain 'Bob' Nairac, who was killed by the Irish Republican Army (IRA) in Northern Ireland during May 1977. Although not a member of the SAS, Nairac did live and work out of the same location as the SAS in Bessbrook, Armagh. He was from a unit known as No 14 Intelligence and Security, a covert unit used to gather information on which the SAS could operate. For some reason best known to himself, Nairac had taken to speaking with an Irish accent, but while this accent was good it was not good enough. He also had the idea that he could pass himself off as an Irishman, and to some degree he did. Two days before his death he had gone to a shop that sold Republican song sheets, and purchased several of the well known melodies. For the next couple of days he relentlessly practised singing the verses until he was word perfect.

A highly intelligent man, and certainly one who was enormously courageous, Nairac decided to visit a local pub close to the border with the Irish Republic. The problem was he had neglected to tell anyone where he was going. He drove off from Bessbrook barracks around 19:30 hours, but he did not reach the

Three Steps Inn, near the village of Dromintee until around 22:00 hours. The pub was in full swing, and at first Bob managed to fit in well with the locals, all having a good time. Several people did a turn on the makeshift stage, mostly singing rebel songs. It was at this point that Bob got up and joined in, putting himself in a position where he would be noticed. It was not so much his accent that gave him away, it was more the fact that no one knew him.

As he attempted to leave and make his way back to the car park, he was followed. Casually, several men inquired as to his identify. A fist fight ensued (Nairac was an excellent boxer and could take care of himself) but during the tussle his 9mm Browning High Power pistol fell to the ground. His assailants grabbed it and he was soon overpowered. What happened next is not clear, but from various people later brought into the police barracks for questioning a rough picture emerged.

While Nairac was held, a telephone call was made to members of the IRA just over the border in the Irish Republic. Blindfolded and gagged, Nairac was taken by car to a field on the border, where the IRA members took control. He was tortured and interrogated in the corner of a field. The main instrument was a fence post, with which they beat him repeatedly around the body and head. Despite what has been called a murderous beating by the IRA themselves, Nairac did not talk. In the end, they shot him with his own pistol and his body was disposed of, never to be recovered. His failure to report his movements made it impossible for him to be rescued.

(Author's Note: Some years later, I was privy to information as to the disposal of his body. It was so horrifying that it has never been made public. Strangely enough, the IRA man relating the story had nothing but pride for the way in which Nairac had suffered in silence.)

Other elite units have also had their failures. US Army Colonel Charles Beckwith

Below: Charles Beckwith, an American colonel who served with the SAS who was instrumental in setting up Delta Force.

Above: Aldo Morro's assassinated bodyguard. Right: An American AC-130H gunship, similar to the ones earmarked for 'Eagle Claw'.

served with the SAS between 1962 and 1964. On his return to the USA, he sought to create an organisation based along similar lines. After many years of trying, Beckwith was given the required authorisation after the GSG 9 and the SAS had carried out the Mogadishu raid (see Hostage-Rescue, pp54-79). His unit was called Delta Force. Two years later it was called upon to carryout a rescue attempt on the US Embassy in Tehran.

After a fundamentalist Moslem coup in Iran, the Shah was toppled from power and his once proud army replaced with a rabble known as the Iranian Revolutionary Guards. On 4 November 1979, the militants took over the US Embassy in Tehran and seized more than 100 Americans hostage. From the earliest days of the crisis, one of the options under constant review and development was a military rescue, although both diplomatic and military endeavours were constantly bedevilled by the continuing chaos in Iran. One of the main problems was the remoteness of Tehran from any available US bases.

The operation was codenamed 'Eagle Claw'. The basic outline plan was to insert Delta Force into Iran covertly, using an old air

base at Masirah Island, off Oman, and from the aircraft carrier *Nimitz*, which was sailing in the Gulf of Oman. The first part of the plan was to establish a firm base – Desert One – about 320km (200 miles) south of Tehran. Delta Force would fly from Masirah to Desert One by Lockheed C-130 Hercules transport aircraft, of which three would carry fuel and another three Delta Force, plus a section of US Army Rangers who would supply protection. Shortly after they had landed at Desert One, eight Sikorsky RH-53D Sea Stallion helicopters would take off from the aircraft carrier and join them. Once the helicopters had arrived at Desert One, they would refuel and carry Delta Force to a hide area just outside Tehran, while the C-130s would return to Masirah. Delta Force was then to lay low for the day just outside the city, ready to attack during the following night (the US Department of Defense had inserted special agents to assist with the vehicles that would transport Delta Force to the city). The helicopters, once they had dropped off Delta Force, would then fly south and hide in the mountains until required for pick-up.

That night, with the assistance of the DoD agents, Delta Force would covertly assault the US Embassy building and rescue the hostages, who would then be extracted by helicopter. Any resistance by the Iranian Revolutionary Guards was to be suppressed, and to stop any supporting Iranian troops coming to the US

Embassy building helicopter gunships were on standby. As the operation was going on around the US Embassy building, a force of Rangers would fly in and secure an airfield called Manzariyeh, some 50km (31 miles) south of Tehran, from where everyone would be lifted out by a Lockheed C-141 StarLifter transport.

The plan involved and relied on many aircraft, and as a precaution special SAR (Search And Rescue) teams were put on standby for use should one of them go down.

The daring plan had presidential approval, and commenced late on the evening of 24 April. Before the mission was launched, Beckwith received a message from Ulrich Wegener, commander of the West German GSG 9 unit, that the Germans were about to insert a TV camera crew into Iran with the intention of covering the American hostage situation, and asking if there was any way he could help. The author believes that Beckwith would have jumped at this, but headquarters shot him down on the grounds that this was to be purely an American operation.

One of the Rangers fired a LAW anti-tank rocket at the tanker

Problems started a few minutes after the C-130s carrying Delta Force touched down at Desert One. A Mercedes bus travelling along a nearby road was stopped by the rescue force and found to have about 30 passengers on board. As these Iranians were being held hostage, a petrol tanker also came along, in this instance from the opposite direction, and for some reason one of the Rangers providing the security fired a LAW anti-tank rocket at the tanker. The tanker burst into flames.

At the same time the driver of a smaller truck, which was coming up behind the tanker, saw both the fire and the soldiers. Before the Americans could do anything, the driver of the tanker leapt from the burning vehicle and jumped into the smaller truck, which made off at speed. Despite the additional risk that the escaped witness posed to the operation, Beckwith decided to press on with the plan, and shortly after this all six C-130s arrived. After dropping off their troops three of them returned to Sharjah. All the Americans now had to do was detain any Iranian civilians until the helicopters (due in 30 minutes) arrived. During this waiting period, kit and equipment were

Left: British and Italian special forces go through their hostage-rescue drills in an unspecified location 'somewhere in England.'

camouflaged and a satellite communications radio link was established.

Then things started to go badly wrong. The choppers were late, and daylight was fast approaching. To make matters worse only six of the choppers arrived, an hour and a half late, as they had hit a bad sandstorm. As the troops started to load, one of the choppers became unserviceable. Beckwith needed all six and almost begged the pilot to fly, but it was no go and the mission was aborted.

Instantly the whole area was illuminated by the fireball as the ammunition ignited

Delta Force's personnel were about to get back into the remaining three C-130s, but as it was lifting off one of the choppers crashed into one of the C-130s. Instantly the whole area was illuminated by the fireball as the ammunition and rockets carried by the C-130 ignited like some giant fireworks display. Five US Air Force aircrew in the C-130 and three US Marines in the RH-53 died. Beckwith put all his men and the chopper pilots on the remaining C-130s and flew back to Masirah, and a jet strike was called in to destroy the choppers that had been left behind. After all the months of planning, for such a disaster to happen was beyond belief, and Beckwith was in tears. It should be pointed out that this failure was not the fault of Delta Force, though many looked in that direction. A few days after Delta Force had returned to the USA, President Carter paid it a visit and had nothing but praise for the unit.

For the SAS in the Gulf War, where much of its men's time was spent behind enemy lines, several stories emerged of daring deeds. Most of what has been written concentrates on 'Bravo Two Zero', but regrettably there were several other incidents in which things went wrong. One such incident occurred when an SAS fighting column spotted two Iraqi trucks laden with missiles. The column was on the move when the trucks suddenly appeared. Both sides spotted each other at the same time, the Iraqi soldiers staring in disbelief at the SAS fighting column. As the trucks rolled past, the SAS decided to give chase. As one SAS member put it: 'It looked like something out of an old

Right: Dutch Marines in action. They have close ties with the SAS, particularly in the sphere of counter-terrorism.

Western, with the Indians chasing the train.' As the two groups closed, the Iraqis came to a sudden stop and the drivers made a run for it. While the bikes swept out and closed with the Iraqi trucks, several of the Land Rovers lined up to fire. The firepower was devastating, the whole area being crisscrossed with tracer fire and grenades from the Mk 19 launchers. It was during this fierce battle that one of the SAS men was killed in the crossfire and another wounded

after his flak vest took the brunt of the impact but did not totally protect him.

Protecting VIPs has also had its failures. In March 1978, for example, the Italian Christian Democrat, Aldo Moro, was kidnapped by the Red Brigade political terrorist group. The attack on his car and escort was carried out in a busy street with military precision. Moro's car was blocked and several automatic weapons opened fire – two of Moro's bodyguards were killed at the scene. Moro himself was whisked away and held for 54 days, during which time the Italian government refused to give in to the terrorists' demands. Eventually Moro's body, shot through the head, was found in the boot of a car parked in a Rome street on 10 May.

During Moro's incarceration by the Red Brigade, a nationwide hunt was launched by the

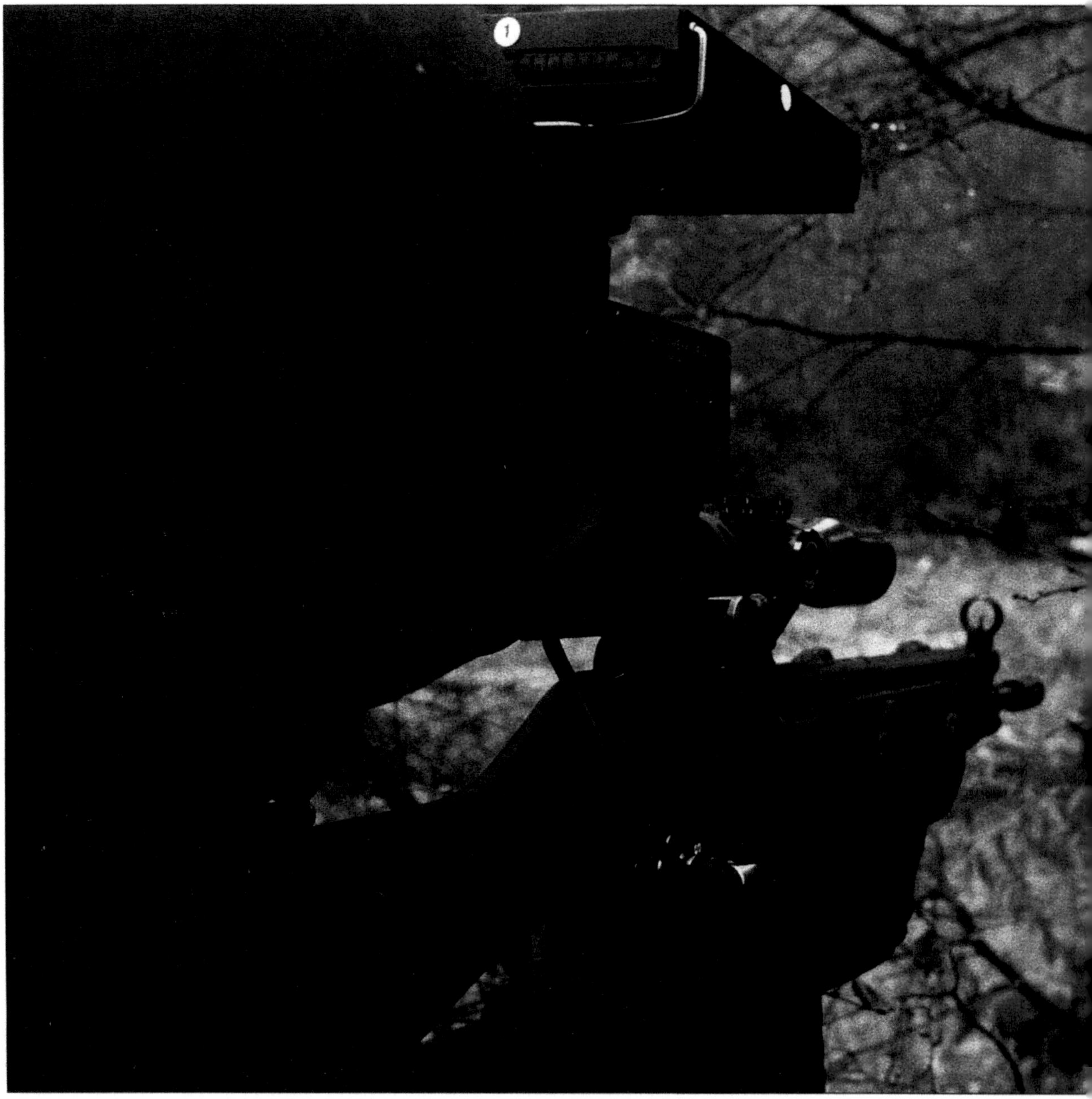

Italian authorities. Many activists were arrested and safe-houses raided, resulting in the seizure of arms and explosives, but nothing that would lead to Moro was uncovered. Interestingly enough, during Moro's kidnapping a four-man SAS team just happened to be in Italy. Their purpose was to infiltrate the Red Brigade with the intention of gathering information. For obvious security purposes, details of the operation can not be disclosed here, but there is one factor that can be tackled – the lack of communication.

After about five days, the four SAS men found themselves in a small hut on the side of a small tree-covered hill. The hut was a safe-house used by the Red Brigade, and although the group travelled only by night, on this occasion the dawn had been light enough for them to be spotted. The vigilant villager, thinking it suspicious that men should be walking around the woods at this hour, informed the police. By 10:00 hours, having eaten breakfast, the four men had settled down to get some sleep. It was at this time that a fat, sweating *Carabiniere* policeman burst through the door, pistol in hand. The SAS men leapt to their feet. The policeman, though armed, could only mutter, 'Good morning, what are you doing here?', when faced by the four.

Without warning several submachine guns opened fire

Not understanding a word of Italian, the lads just shrugged their shoulders and spoke in English. The policeman then made a run for the door. The SAS guys, now laughing, watched as he ran down the hill, then went back to sleep.

Some two hours later, one of them needed to relieve himself and so ventured outside to the nearest bush. Seconds later he burst back into the room with the news that there were lots of armed police surrounding the hut. The guys all went outside for a look. Without warning several submachine guns opened fire. As they did, the SAS group made a break for it, running like the wind until they were clear of any danger. They later re-established contact with the authorities, and were soon placed in a safe-house prior to their departure back to England. The problem had been lack of communication between the Italian intelligence groups. Poor communications can be rectified, but often rescues fail simply because they run out of luck.

Left: A NOCS sniper takes aim during a VIP rescue exercise. Only constant training can reduce the possibility of failure.

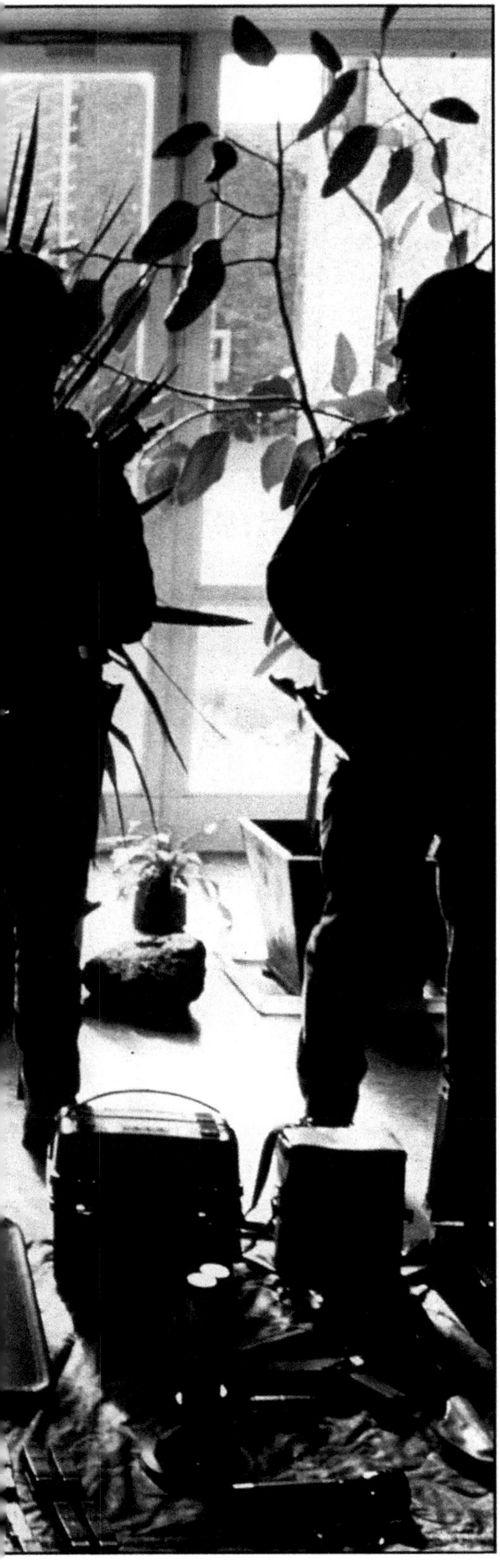

WEAPONS AND EQUIPMENT

Successful SAS rescue operations need first and foremost highly trained soldiers. In turn, those soldiers need reliable hardware to give them the edge against their opponents.

The skills of the SAS soldier have been clearly described in the preceding chapters, but in the realms of war and violence, the one thing every soldier needs apart from his own skills is the right equipment. For the SAS, much of the equipment is straightforward British Army issue, but for more complicated tasks, such as those undertaken by the anti-terrorist team, specialist equipment is required. When it comes to equipment choice, there are three different roles to consider. First, the straightforward infantry deployment, such as in Borneo, Oman, the Falklands and the Gulf War. Second, there is the covert, plainclothes undercover role, such as in Northern Ireland and various bodyguard assignments around the world. Finally, and probably the role which demands the most specialised equipment, is the anti-terrorist team organisation.

The SAS soldier wears and carries much the same as his British Army infantry counterpart, although specialist

Left: A GIGN team with just some of the weapons and equipment needed for hostage-rescue operations, including the MP5 submachine gun.

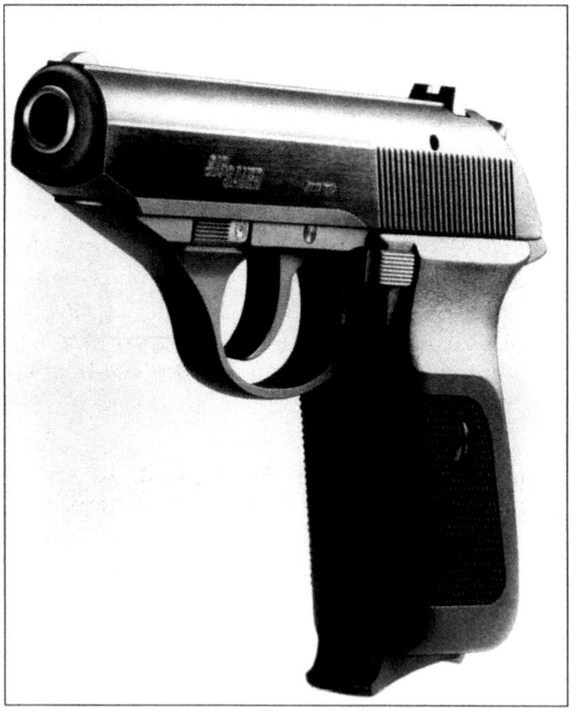

Above: The SIG-Sauer P230 pistol, a compact weapon designed for personal defence. It is ideal for SAS covert missions.

equipment will depend on the task undertaken. Normally, most SAS soldiers operate deep behind enemy lines or in isolated areas like the jungle. Such an environment demands that the soldier must carry everything he is likely to need. Before operations take place, therefore, SAS soldiers spread out their equipment for a rigorous examination, making sure that nothing is duplicated or omitted. But first they attend to what they are wearing.

In the combat role the normal dress is the SAS windproof smock, windproof trousers and a woollen hat or cap. A Norwegian shirt is normally worn, but the number of additional layers beneath this depends on the climate. His equipment is divided into two parts: his belt kit and his bergen. The bergen contains everything else the soldier needs: secondary ammunition, explosives, water, rations, sleeping bag and sleeping mat. If the signaller has to dump his bergen, the radio set still has to be taken and is therefore always in a separate bag attached to the top of the bergen. The SAS belt kit contains magazines for immediate use, water, medical pack and an escape-and-evasion kit. The last is carried at all times, and contains enough items for the soldier to make it back safely to his own lines. Items such as search and rescue beacons are also carried on the belt kit. In a bug-out situation, where the patrol has been hit hard by the enemy, bergens are discarded and the soldier quickly runs away with only his belt kit around his waist. He will also have his assault rifle in his hands.

When it comes to personal weapons SAS soldiers favour the American M16. The Colt Armalite has been in service with the SAS since the late 1960s. The AR-15 early model had many problems, most of which involved a jammed cartridge case in the breech. In Oman in the 1970s, it was not uncommon to see a small length of welding rod attached to the side of SAS AR-15's. Poking this down the barrel was the simplest and quickest method of clearing the stoppage. On the plus side, both the weapon and the ammunition were, and are, light to carry. It also had low recoil, making it comfortable to fire.

All in all, the M16 is a weapon used and liked by the SAS

The latest version of the M16 to be used by the SAS is the M16A2, which has several significant modifications. It has two firing options (semi- and full-automatic) and a three-round burst capability. It also fires NATO-approved 5.56mm ammunition (the AR-15 would also fire NATO ammunition, but in Oman it was found that this drastically increased the number of stoppages, and as a direct result ammunition was shipped in from the USA). Like the earlier version, the weapon is easy to fire and, despite some early misgivings, is amazingly accurate. It feeds from a 20- or 30-round magazine, although in the early days the 30-round magazine was not readily available. All in all, the M16 is a weapon used and liked by the SAS.

The M16's firepower can be substantially increased with the addition of an M203 grenade launcher, which can be fitted underneath the barrel of the M16A2. A folding quadrant sight is fitted to the left side of the carrying handle, and is graded from 40-400m (132-1320ft). The 40mm (1.5in) grenade comes in a variety of role-optimised forms, but to the author's knowledge the SAS has only ever used high explosive. The M203 has proved its worth in both the jungle and the desert.

SAS foot patrols are vulnerable to enemy armoured vehicles. They therefore need anti-armour weapons which are lightweight. The 66mm (2.6in) Light Anti-tank Weapon (LAW) has been used by the SAS since the early 1970s. Although designed for use against armoured fighting vehicles, it has proved to be very effective at dislodging any enemy revealing a determination to stay put. The weapon is small and very light – 2.36kg (5.19lb) – and several can be carried by one soldier. Once fired, the launcher unit is discarded. It has an effective range of 100-300m (330-990ft), but will travel to a maximum range of almost 1000m (3285ft). Operation is simple, with pictorial instructions printed on the side. When the two transit lids have been removed, the rear section, containing the rocket, can be extended. This causes the front and rear sights to pop up ready for use. Arming is effected by pulling the small safety catch forward. The M72 has since been

Below: The SIG-Sauer P226 pistol, which can accommodate a 15- or 20-round magazine.

improved into what is designated the 'E' series. Additionally, the British Army has produced its own LAW 80 with the same basic concept.

Because SAS teams often operate at night, they need excellent night sights for their weapons. The Regiment uses two types of passive night sight: one unit that is fitted to weapons, and one that is used for close target reconnaissances. SAS anti-terrorist snipers, for example, each have at least one weapon permanently fitted with a night sight and zeroed for night assaults. On the other hand, small hand-held pocketscopes are used for surveillance in Northern Ireland.

A new type of targetting aid that is fairly new to the Regiment is the Type SS82 laser designator. It is a sophisticated piece of equipment that fits in well with the behind-the-lines patrolling that is a feature of the SAS art of war. Once an enemy target has been positively identified, it is possible for the patrol to set up the laser sight. The sight itself provides

Left: Light, accurate, deadly. The M72 anti-tank weapon is ideally suited to different types of SAS operation.

an invisible beam that is directed at the target. The beam then reflects off the target in a scattered radiation pattern that can be detected by laser-seeking weapons, which then lock onto the target. This system is excellent for forward air control as it allows the attacking aircraft to approach at low level and at high speed. The pilot is not required to see the target himself, which makes a single-pass attack possible and thereby reduces exposure to enemy ground defences. From the SAS point of view, it means that some targets can be destroyed without the need to be on-site or carry huge amounts of explosives around.

The Regiment also mounts vehicle patrols, which means more equipment and ammunition can be carried. The vehicles themselves have to be hardy and reliable for long-range operations, and so it is perhaps inevitable that the Regiment uses Land Rovers. Anyone who lives in and around Hereford will know about the pink cut-down Land Rovers that are often seen on the roads. Why pink? An aircraft shot down during World War II was discovered some years later, and it was found that the desert sun had scorched it pink, thereby causing it to blend into the sand (any desert soldier knows that during a certain light, the sand is very pink).

All Land Rovers are fitted with front and rear smoke dischargers

Land Rovers have been in service with the SAS since the 1950s. Most are fitted with long-range tanks and carry a variety of weaponry. Normally, a General Purpose Machine Gun (GPMG) is fitted ahead of the front passenger position. The main weapon is fitted in the rear, and this can vary from a second GPMG, a Mk 19 grenade launcher or a 7.62mm Minigun multi-barrel machine gun, depending on the required role. All Land Rovers are fitted with front and rear smoke dischargers for use when the vehicle runs into trouble, and these can lay down an instant smoke screen. Personal weapons for the crew are kept in metal sleeves attached to each side of the vehicle, giving instant access.

For a while the Light Strike Vehicle (LSV), an upgraded beach-buggy, was very popular with special forces units including the SAS. The first LSVs trialled by the Regiment were excellent and, given the chance to prove their efficiency, would have become part of Mobility Troop's equipment despite their resemblance to something out of a *Mad Max* film. They are simple, fast and can carry a wide variety of weapons, but unfortunately they weren't up to the job. The SAS used LSVs before the Gulf War, but during the first weeks of the conflict the vehicles had so many problems that they were 'binned'. LSVs lack the Land Rover's payload and range, and also cannot carry the firepower and extra ammunition that the Land Rover 100 can support.

The SAS has two anti-terrorist teams, each with four Range Rovers for use by the team commander and the assault groups. Three vans

are used by the snipers; these vans also carry the explosive devices and specialist communication equipment. As the Range Rovers are full of personal weapons and ammunition, a Volvo articulated truck is used for the heavier equipment. This carries the platforms for the Range Rovers and all ladder equipment that the team might use. It is possible to attach the ladders to the platforms for extra height. All other bulky equipment, such as boats, main stores and various heavy specialist items, are carried by a large truck.

What of hostage-rescue clothing? Helmets are either Kevlar for ballistic protection – when someone is shooting at you – or Protec for general head protection: falling off a ladder during training, for example. Underwear, balaclavas and overalls are all of flame-resistant material. Assault boots are either Adidas or Puma, though in certain situations Hi-Tec trainers are worn. Gloves are heavy duty for fast roping, but in a normal assault, light pilot-type gloves are worn. Bristol Armour still supplies the best body armour, which can be fitted with Kevlar plates front and rear.

Unlike most other submachine guns, the MP5 fires from a closed bolt

The sniper has a special gilly suit and heavy thermal underwear, although he also carries full assault clothing and equipment, and this allows his role to be switched as required.

Hostage-rescue units require weapons that work first time every time – the first few seconds of an asault are crucial to the outcome. The SAS therefore uses one of the most reliable and accurate submachine guns in the world: the 9mm Heckler & Koch MP5. Heckler & Koch was founded in 1947 by three former Mauser employees. Originally it did not manufacture weapons, but by 1959 it had returned to the trade it knew best. The company had its first big break in 1959, when its G3 assault rifle was adopted by the West German army. The MP5 was developed from this successful weapon and shares many of the same characteristics, including weight. When the weapon was first introduced it was used by the German border police. Unlike most other submachine guns, the MP5 fires from a closed and locked bolt, using

Left: A group of SAS troops in Scotland undergoing training in the use of the laser target designator.

the same delayed blowback action as the G3. This makes the weapon expensive to produce, but this is compensated for by the increased safety and accuracy. The MP5 is overwhelmingly, and justly, the chosen weapon of the world's anti-terrorist units. This weapon is precision engineering at its very best, and the perfect fit of the parts is reflected in its handling as well as its stripping and reassembly.

Close behind comes the Browning High Power pistol. This 9mm pistol is extremely popular with the SAS, mainly for its reliability. It is the one weapon that every member of the SAS can strip and assemble in his sleep, with his eyes closed while doing backwards somersaults! That said, all troopers continually practise how to clear every type of stoppage, move with the weapon, shoot from any given angle and learn the double tap (two shots fired in quick succession). Although it is a somewhat old design, the High Power holds 13 hard-hitting rounds. The Browning pistol has been used by the SAS from the back streets of Aden in the early 1960s to the 1991 Gulf War, but it looks

as if it is finally being eclipsed. The Regiment has recently switched to the 9mm SIG-Sauer P 226 pistol as a secondary weapon.

This 9mm double-action, locked-breech automatic is one of the best pistols available. A combination of Swiss and German engineering technology, it is superbly accurate and is very reliable. Its finest feature is the 15-round magazine – very important in a firefight.

Pistols are excellent for close-quarter work, but for long-range shooting the Regiment requires sniper rifles. The current sniper weapon used by the SAS is the 7.62mm Accuracy International PM. The PM is a bolt-action rifle with a stainless steel barrel which provides head-shot accuracy at ranges between 300-600m (990-1965ft). The rifle itself is very comfortable to fire, and the sniper's head need not be moved during bolt operation, allowing for continuous observation. It has a bipod and a retractable spike on the rear of the butt. The box magazine holds 12 rounds, with aiming through a standard Schmitt and Bender telescopic sight.

For short-range sniping the SAS uses a Heckler & Koch G3SG. This 7.62mm self-loading sniper rifle is used for targets in the range bracket of 200-300m (660-990ft). The weapon is basically a G3 rifle with an upgraded barrel, a variable trigger and a comfortable butt. It is excellent for multiple targets, or rapid long-range shooting.

For hostage-rescue operations the Regiment also uses stun grenades

At the other end of the scale, the SAS uses Winchester and Remington pump-action shotguns. They are normally used by the anti-terrorist team to blow hinges off doors. Solid lead ball is the most popular round, although gas and buckshot are also carried by the team.

For hostage-rescue operations the Regiment also uses stun grenades. Royal Enfield Ordnance, at the request of the SAS, experimented with various devices that would give the assaulting troops vital seconds to come to grips with the terrorists. One that seemed most favourable was the stun grenade. The device is a G60, which produces a loud noise (160dB) combined with a blinding light (300,000cd) without any harmful fragmentation. This is capable of stunning anyone in close proximity for a period of 3-5 seconds when detonated, and is one of the most effective items in the anti-terrorist armoury. The effect is a million times more effective than the strobes in a disco.

Left: The Longline Light Strike Vehicle, a piece of equipment that failed to make the grade with the Regiment.

Other specialist equipment used by the hostage-rescue teams are passive night goggles such as the Ground Owl, which is a sixth-generation imaging system that provides excellent night vision. Snipers also use laser rangefinders capable of giving an accurate range readout of anything up to a maximum of 9km (5.6 miles).

The Range Rover has been in service with the SAS anti-terrorist team from day one and it is still used. The first vehicles were straightforward production models, but over the years many modifications have been made. The first one concerned safety. When the team was deployed in the 1970s, it would race to the terrorist scene with a police escort, sirens sounding and blue lights flashing to clear the route. During these fast drives many of the Range Rovers crashed. Although several Range Rovers were totally written off, the occupants miraculously escaped with minor cuts and bruises. Roll-bars were therefore fitted, and the procedure for getting to an incident has acquired a more thoughtful approach. Communications in these vehicles is second to none, enabling commanders to talk securely to just about anyone in the world if they wish. A new add-on feature to the Range Rovers are the platforms. These are designed to carry the assault teams directly to the objective at high speed. They can be fitted to the bonnet, sides or roof and can carry up to four men.

To get into buildings, trains or aircraft during a hostage-rescue requires entry equipment. One of the primary tools of this type of work is the thermic lance. This is basically a flexible plastic tube filled with magnesium granules, with a valve to allow a mixture of gases to be fed through the tube at regulated pressures. The result is a flame of very high temperature created as the magnesium and gases burn. This heat is sufficient to cut metal by melting it, and no thickness is a problem for the thermic lance. In normal usage, the unit fits into a small briefcase which, when opened, reveals a length of magnesium tubing, about 3m (10ft) long, a control panel and, beneath this control panel, three small cylinders of gas. The case also has an attachment allowing a larger gas bottle to be fitted for an extension of the burn time during training. Improvements have been made over the years, but some of the early models had faults. During one training period at the demolition bunker, for example, several SAS men from the anti-terrorist team narrowly escaped being blown to pieces. As they happily cut their way through a sheet of metal, one of them noticed that the case had flames coming from it. He gave a quick warning and all three ran for safety. No sooner had they reached the bunker door when the master cylinder, from which they had been using, exploded in a massive fire ball.

It is a device of fearsome appearance resembling an old-type mortar

The Harvey Wallbanger is an air-compression device used for wall entries. The kit was designed to replace the frame charge where use of the latter would put any hostages at risk. It is a device of fearsome appearance resembling an old-type heavy mortar, but it works. The barrel is loaded with a large plastic bottle filled with water (the quantity can be varied to create various pressures when it hits the wall). This plastic bullet is fired from the Harvey Wallbanger by compressed air (like the water, the pressure can be regulated). If the wall requires it, a mixture of water and concrete can be used, with devastating effect. The result of a round hitting a cavity wall (internal block with outer brick) is devastating. The hole created by two Harvey Wallbangers, placed one on top of the other, is sufficient for an SAS man in full assault gear to climb through. The device can also be fired up at a ceiling from the floor below. In such a case the device is only fired with sufficient power to shake the floor, resulting in everyone in the room above being thrown to the floor in a kind of mini-earthquake. This gives the entry team a precious few seconds to gain entry to the room and neutralise the terrorists.

Left: The 9mm Browning High Power pistol, a weapon that is particularly reliable and robust. Overleaf: Hostage-rescue assault ladders.

Index

#0019 - 141118 - C0 - 240/189/10 - PB - 9781782747529